being digital

being

digital

NICHOLAS NEGROPONTE

ALFRED A. KNOPF   NEW YORK   1995

This Is a Borzoi Book
Published by Alfred A. Knopf, Inc.
Copyright © 1995 by Nicholas P. Negroponte
All rights reserved under International and Pan-American
Copyright Conventions. Published in the United States by
Alfred A. Knopf, Inc., New York, and simultaneously in
Canada by Random House of Canada Limited, Toronto.
Distributed by Random House, Inc., New York.
Portions of this work were originally published in slightly
different form in *Wired* magazine.

Library of Congress Cataloging-in-Publication Data
Negroponte. Nicholas.
Being digital / Nicholas Negroponte. — 1st ed.
p.   cm.
Includes index.
ISBN 0-679-43919-6
1. Digital communications—Social aspects.   2. Technology and
civilization.   3. Computer networks—Social aspects.   4. Interactive
media—Social aspects.   I. Title.
TK5103.7.N43   1995
303.48'33—dc20   94-45971 CIP

Published February 2, 1995
Reprinted Twice
Fourth Printing, April 1995

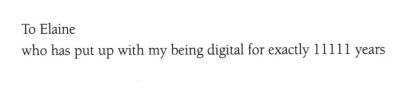

To Elaine
who has put up with my being digital for exactly 11111 years

CONTENTS

being digital

INTRODUCTION:
THE PARADOX OF A BOOK

Being dyslexic, I don't like to read. As a child I read train timetables instead of the classics, and delighted in making imaginary perfect connections from one obscure town in Europe to another. This fascination gave me an excellent grasp of European geography.

Thirty years later, as director of the MIT Media Lab, I found myself in the middle of a heated national debate about the transfer of technology from U.S. research universities to foreign companies. I was soon summoned to two industry-government meetings, one in Florida and one in California.

At both meetings, Evian water was served in one-liter glass bottles. Unlike most of the participants, I knew exactly where Evian was from my timetables. Evian, France, is more than five hundred miles from the Atlantic Ocean. Those heavy glass bottles had to traverse almost one-third of Europe, cross the Atlantic, and, in the case of California, travel an additional three thousand miles.

So here we were discussing the protection of the American computer industry and our electronic competitiveness, when we seemingly could not even provide American water at an American conference.

Today, I see my Evian story not so much being about French mineral water versus American, but illustrating the fundamental difference between atoms and bits. World trade has traditionally consisted of exchanging atoms. In the case of Evian water, we were shipping a large, heavy, and inert mass, slowly, painfully, and expensively, across thousands of miles, over a period of many days. When you go through customs you declare your atoms, not your bits. Even digitally recorded music is distributed on plastic CDs, with huge packaging, shipping, and inventory costs.

This is changing rapidly. The methodical movement of recorded music as pieces of plastic, like the slow human handling of most information in the form of books, magazines, newspapers, and videocassettes, is about to become the instantaneous and inexpensive transfer of electronic data that move at the speed of light. In this form, the information can become universally accessible. Thomas Jefferson advanced the concept of libraries and the right to check out a book free of charge. But this great forefather never considered the likelihood that 20 million people might access a digital library electronically and withdraw its contents at no cost.

The change from atoms to bits is irrevocable and unstoppable.

Why now? Because the change is also exponential—small differences of yesterday can have suddenly shocking consequences tomorrow.

Did you ever know the childhood conundrum of working for a penny a day for a month, but doubling your salary each day? If you started this wonderful pay scheme on New Year's Day, you would be earning more than $10 million per day on the last day of January. This is the part most people remember. What we do not realize is that, using the same scheme, we would earn only about $1.3 million if January were three days shorter (i.e., February). Put another way, your cumulative income for that whole month of February would be roughly $2.6 million, instead of the $21 million you earned in total during January. When an effect is exponential, those last three days mean a lot! We are approaching those last three days in the spread of computing and digital telecommunications.

In the same exponential fashion, computers are moving into our daily lives: 35 percent of American families and 50 percent of American teenagers have a personal computer at home; 30 million people are estimated to be on the Internet; 65 percent of new computers sold worldwide in 1994 were for the home; and 90 percent of those to be sold this year are expected to have modems or CD-ROM drives. These numbers do not even include the fifty microprocessors in the average 1995 automobile, or the microprocessors in your toaster, thermostat, answering machine, CD player, and greeting cards. And if I am wrong about any of the numbers above, just wait a minute.

And the rate at which these numbers are growing is astonishing. The use of one computer program, a browser for the Internet called Mosaic, grew 11 percent per week between February and

December 1993. The population of the Internet itself is now increasing at 10 percent per month. If this rate of growth were to continue (quite impossibly), the total number of Internet users would exceed the population of the world by 2003.

Some people worry about the social divide between the information-rich and the information-poor, the haves and the have-nots, the First and the Third Worlds. But the real cultural divide is going to be generational. When I meet an adult who tells me he has discovered CD-ROM, I can guess that he has a child between five and ten years old. When I meet someone who tells me she has discovered America Online, there is probably a teenager in her house. One is an electronic book, the other a socializing medium. Both are being taken for granted by children the same way adults don't think about air (until it is missing).

Computing is not about computers any more. It is about living. The giant central computer, the so-called mainframe, has been almost universally replaced by personal computers. We have seen computers move out of giant air-conditioned rooms into closets, then onto desktops, and now into our laps and pockets. But this is not the end.

Early in the next millennium your right and left cuff links or earrings may communicate with each other by low-orbiting satellites and have more computer power than your present PC. Your telephone won't ring indiscriminately; it will receive, sort, and perhaps respond to your incoming calls like a well-trained English butler. Mass media will be redefined by systems for transmitting and receiving personalized information and entertainment. Schools will change to become more like museums and playgrounds for children to assemble ideas and socialize with other children all over the world. The digital planet will look and feel like the head of a pin.

As we interconnect ourselves, many of the values of a nation-state will give way to those of both larger and smaller electronic communities. We will socialize in digital neighborhoods in which physical space will be irrelevant and time will play a different role. Twenty years from now, when you look out a window, what you see may be five thousand miles and six time zones away. When you watch an hour of television, it may have been delivered to your home in less than a second. Reading about Patagonia can include the sensory experience of going there. A book by William Buckley can be a conversation with him.

So why an old-fashioned book, Negroponte, especially one without a single illustration? Why is Knopf shipping *Being Digital* as atoms instead of bits, when these pages, unlike Evian water, can be so easily rendered into digital form, from whence they came? There are three reasons.

First, there are just not enough digital media in the hands of executives, politicians, parents, and all those who most need to understand this radically new culture. Even where computers are omnipresent, the current interface is primitive—clumsy at best, and hardly something with which you might wish to curl up in bed.

A second reason is my monthly column in *Wired* magazine. The rapid and astonishing success of *Wired* has shown that there is a large audience for information about digital life-styles and people, not just theory and equipment. I received so much thoughtful feedback from my (text only) column that I decided to repurpose many of the early themes, because a great deal has changed even in the short time since those stories were written. And that is what they are: stories drawn from years of inventing new systems for computer graphics, human communications, and interactive multimedia.

The third is a more personal, slightly ascetic reason. Interactive multimedia leaves very little to the imagination. Like a Hollywood film, multimedia narrative includes such specific representations that less and less is left to the mind's eye. By contrast, the written word sparks images and evokes metaphors that get much of their meaning from the reader's imagination and experiences. When you read a novel, much of the color, sound, and motion come from you. I think the same kind of personal extension is needed to feel and understand what "being digital" might mean to your life.

You are expected to read yourself into this book. And I say this as somebody who does not like to read.

bits are bits

THE DNA
OF INFORMATION

1

BITS AND ATOMS

The best way to appreciate the merits and consequences of *being digital* is to reflect on the difference between bits and atoms. While we are undoubtedly in an information age, most information is delivered to us in the form of atoms: newspapers, magazines, and books (like this one). Our economy may be moving toward an information economy, but we measure trade and we write our balance sheets with atoms in mind. GATT is about atoms.

I recently visited the headquarters of one of America's top five integrated circuit manufacturers. I was asked to sign in and, in the process, was asked whether I had a laptop computer with me. Of course I did. The receptionist asked for the model and serial number and for its value. "Roughly, between one and two million dollars," I said. "Oh, that cannot be, sir," she replied. "What do you mean? Let me see it." I showed her my old Power-Book and she estimated its value at $2,000. She wrote down that amount and I was allowed to enter the premises. The point is that while the atoms were not worth that much, the bits were almost priceless.

Not long ago I attended a management retreat for senior executives of PolyGram in Vancouver, British Columbia. The purpose was to enhance communications among senior management and to give everybody an overview of the year to come, including many samples of soon-to-be-released music, movies, games, and rock videos. These samples were to be shipped by FedEx to the meeting in the form of CDs, video-cassettes, and CD-ROMs, physical material in real packages that have weight and size. By misfortune, some of the material was held up in customs. That same day, I had been in my hotel room shipping bits back and forth over the Internet, to and from MIT and elsewhere in the world. My bits, unlike Poly-Gram's atoms, were not caught in customs.

The information superhighway is about the global movement of weightless bits at the speed of light. As one industry after another looks at itself in the mirror and asks about its future in a digital world, that future is driven almost 100 percent by the ability of that company's product or services to be rendered in digital form. If you make cashmere sweaters or Chinese food, it will be a long time before we can convert them to bits. "Beam me up, Scotty" is a wonderful dream, but not likely to come true for several centuries. Until then you will have to rely on FedEx, bicycles, and sneakers to get your atoms from one place to another. This is not to say that digital technologies will be of no help in design, manufacturing, marketing, and management of atom-based businesses. I am only saying that the core business won't change and your product won't have bits standing in for atoms.

In the information and entertainment industries, bits and atoms often are confused. Is the publisher of a book in the information delivery business (bits) or in the manufacturing business (atoms)? The historical answer is both, but that will change rapidly as information appliances become more ubiquitous and user-friendly. Right now it is hard, but not impossible, to compete with the qualities of a printed book.

A book has a high-contrast display, is lightweight, easy to "thumb" through, and not very expensive. But getting it to you includes shipping and inventory. In the case of textbooks, 45 percent of the cost is inventory, shipping, and returns. Worse, a book can go out of print. Digital books never go out of print. They are always there.

Other media has even more immediate risk and opportunity. The first entertainment atoms to be displaced and become bits will be those of videocassettes in the rental business, where consumers have the added inconvenience of having to return the atoms and being fined if they are forgotten under a couch ($3 billion of the $12 billion of the U.S. video rental business is said to be late fines). Other media will become digitally driven by the combined forces of convenience, economic imperative, and deregulation. And it will happen fast.

WHAT IS A BIT, ANYHOW?

A bit has no color, size, or weight, and it can travel at the speed of light. It is the smallest atomic element in the DNA of information. It is a state of being: on or off, true or false, up or down, in or out, black or white. For practical purposes we consider a bit to be a 1 or a 0. The meaning of the 1 or the 0 is a separate matter. In the early days of computing, a string of bits most commonly represented numerical information.

Try counting, but skip all the numbers that have anything other than a 1 and a 0 in them. You end up with the following: 1, 10, 11, 100, 101, 110, 111, etc. Those are the respective binary representations for the numbers 1, 2, 3, 4, 5, 6, 7, etc.

Bits have always been the underlying particle of digital computing, but over the past twenty-five years we have greatly expanded our binary vocabulary to include much more than just numbers. We have been able to digitize more and more types of information, like audio and video, rendering them into a similar reduction of 1s and 0s.

Digitizing a signal is to take samples of it, which, if closely spaced, can be used to play back a seemingly perfect replica. In an audio CD, for example, the sound has been sampled 44.1 thousand times a second. The audio waveform (sound pressure level measured as voltage) is recorded as discrete numbers (themselves turned into bits). Those bit strings, when played back 44.1 thousand times a second, provide a continuous-sounding rendition of the original music. The successive and discrete measures are so closely spaced in time that we cannot hear them as a staircase of separate sounds, but experience them as a continuous tone.

The same can be true for a black-and-white photograph. Imagine an electronic camera as laying a fine grid over an image and then recording the level of gray it sees in each cell. If we set the value of black to be 0 and the value of white to be 255, then any gray is somewhere between the two. Conveniently, a string of 8 bits (called a byte) has 256 permutations of 1s and 0s, starting with 00000000 and ending with 11111111. With such fine gradations and with a fine grid, you can perfectly reconstruct the picture for the human eye. As soon as you use a coarser grid or an insufficient number of gray levels, you start to see digital artifacts, like contours and blockiness.

The emergence of continuity from individual pixels is analogous to a similar phenomenon on a much finer scale in the familiar world of matter. Matter is made of atoms. If you could look at a smoothly polished metal surface at a subatomic scale, you would see mostly holes. It appears smooth and solid because the discrete pieces are so small. Likewise digital output.

But the world, as we experience it, is a very analog place. From a macroscopic point of view, it is not digital at all but continuous. Nothing goes suddenly on or off, turns from black to white, or changes from one state to another without going through a transition. This may not be true at microscopic level, where things that we interact with (electrons in a wire or photons in our eye) are discrete. But there are so many of them that we approximate them as continuous. There are, after all, roughly 1,000,000,000,000,000,000,000,000 atoms in this book (a very analog medium).

There are many merits to digitization. Some obvious ones include data compression and error correction, which is important in the delivery of information through a costly or noisy channel. Broadcasters, for example, can save money, and viewers can see or hear studio-quality picture and sound. But we are discovering that the consequences of being digital are far more important than those.

When using bits to describe sound and picture, there is a natural advantage to using as few bits as possible. This is akin to energy conservation. However, the number of bits you devote per second or per square inch relates directly to the fidelity of the music or image. Typically, one has an interest in digitizing at a very high resolution and then using a less-resolved version of the sound or picture for one application or another. For example, a color image might be digitized at very high resolution for final print copy but used at a lower resolution for a computer-aided page layout system. The economy of bits is driven in part by the constraints of the medium on which it is stored or through which it is delivered.

The number of bits that can be transmitted per second through a given channel (like copper wire, radio spectrum, or optical fiber) is the bandwidth of that channel. It is a measure of how many bits can get down a given pipe. That number or capacity needs to be matched carefully with the number of bits needed to render a given type of data (voice, music, video): 64,000 bits per second is more than ample for high-quality voice, 1.2 million bits per second is more than sufficient for high-fidelity music, and 45 million bits per second is terrific for rendering video.

Over the past fifteen years, however, we have learned how to compress the raw digital form of sound and picture by looking at the bits over time, space, or both and removing the intrinsic redundancies and repetitions. In fact, one of the reasons that all media has become digital so quickly is that we achieved very high levels of compression much sooner than most people predicted. In fact, as recently as 1993, some Europeans were arguing that digital video would not be a reality until the next millennium.

Five years ago, most people did not believe you could reduce the 45 million bits per second of raw digital video to 1.2 million bits per second. Yet in 1995 we can compress and decompress, encode and decode, video at that rate, inexpensively and with high quality. It is as if we suddenly have been able to make freeze-dried cappuccino, which is so good that by adding water, it comes back to us as rich and aromatic as any freshly brewed in an Italian café.

WHEN ALL MEDIA ARE BITS

Being digital allows you to deliver a signal with information added to correct errors such as telephone static, radio hiss, or television snow. These artifacts can be removed from the digital signal using a few extra bits and increasingly sophisticated error-correction techniques that are applied to one form of noise or another, in one medium or another. On your audio CD, one-third of the bits are used for error correction. Similar techniques can be applied to existing television so that each home receives studio-quality broadcast—so much clearer than what you get today that you could mistake it for so-called high definition.

Error correction and data compression are the two obvious reasons for digital television. You can put four studio-quality digital TV signals into the same bandwidth that previously accommodated one noisy analog television transmission. You deliver a better picture, and, using the same channel, you potentially get four times the audience share and advertising revenue.

Better and more efficient delivery of what already exists is what most media executives think and talk about in the context of being digital. But like the Trojan horse, the consequence of this gift will be surprising. Wholly new content will emerge from being digital, as will new players, new economic models, and a likely cottage industry of information and entertainment providers.

When all media is digital—because bits are bits—two fundamental and immediate results will be observed.

First, bits commingle effortlessly. They start to get mixed up and can be used and reused together or separately. The mixing of audio, video, and data is called *multimedia*; it sounds complicated, but is nothing more than commingled bits.

Second, a new kind of bit is born—a bit that tells you about the other bits. These new bits are typically "headers," which are well known to newspaper reporters who file "slugs" (which we never see) to identify a story. Such headers are also familiar to scientific authors who are asked to provide key words with their journal articles. These header bits can be a table of contents or a description of the data that follow. On your CD today, you have simple headers that allow you to skip from song to song and, in some cases, get more data about the music. These bits are not visible or audible but tell you, your computer, or a special-purpose entertainment appliance about the signal.

These two phenomena, commingled bits and bits-about-bits, change the media landscape so thoroughly that concepts like video-on-demand and shipping electronic games down your local cable are just trivial applications—the tip of a much more profound iceberg. Think about the consequences of a broadcast television show as data which includes a computer-readable description of itself. You could record based on content, not time of day or channel. Or, what about a single digital description that can generate a program in audio, video, or textual form at the receiver? And if moving these bits around is so effortless, what advantage would the large media companies have over you and me?

Being digital begs such questions. It creates the potential for new content to originate from a whole new combination of sources.

WHERE INTELLIGENCE LIVES

Broadcast television is an example of a medium in which all the intelligence is at the point of origin. The transmitter determines everything and the receiver just takes what it gets. In fact, per cubic inch, your current TV set is perhaps the dumbest appliance in your home (and I'm not even talking about the programs). If you have a microwave oven, it likely has more microprocessors than your TV. Instead of thinking of the next evolutionary step of television as increased resolution, better color, or more programs, think of it as a change in the distribution of intelligence—or, more precisely, the movement of some intelligence, from the transmitter to the receiver.

A newspaper is also produced with all the intelligence in the transmitter. But the medium of large-format paper provides some relief to the "sameness" of information, as it can be consumed differently, by different people, at different times. We browse and flip through pages, guided by headlines and pictures, each of us treating very differently the identical bits delivered to hundreds of thousands of people. The bits are the same, but the reading experience is different.

One way to look at the future of being digital is to ask if the quality of one medium can be transposed to another. Can the television experience be more like the newspaper experience? Many people think of newspapers as having more depth than television news. Must that be so? Similarly, television is considered a richer sensory experience than what newspapers can deliver. Must that be so?

The answer lies in creating computers to filter, sort, prioritize, and manage multimedia on our behalf—computers that read newspapers and look at television for us, and act as editors when we ask them to do so. This kind of intelligence can live in two different places.

It can live at the transmitter and behave as if you had your own staff writers—as if *The New York Times* were publishing a single newspaper tailored to your interests. In this first example, a small subset of bits has been selected especially for you. The bits are filtered, prepared, and delivered to you, perhaps to be printed in the home, perhaps to be viewed more interactively with an electronic display.

The second example is one in which your news-editing system lives in the receiver and *The New York Times* broadcasts a very large number of bits, perhaps five thousand different stories, from which your appliance grabs a select few, depending on your interests, habits, or plans for that day. In this instance, the intelligence is in the receiver, and the dumb transmitter is indiscriminately sending all the bits to everybody.

The future will not be one or the other, but both.

DEBUNKING BANDWIDTH 2

FROM A TRICKLE TO A DOWNPOUR

I n the late 1960s, when I was an assistant professor of computer graphics, nobody knew what that was. Computers were totally outside everyday life. Today, I hear sixty-five-year-old tycoons boasting about how many bytes of memory they have in their Wizards or the capacity of their hard disks. Some people talk half-knowingly about the speed of their computer (thanks to the brilliant campaign of "Intel Inside") and affectionately (or not) about the flavor of their operating systems. I recently met one socialite, a wealthy and charming woman, who knew so much about Microsoft's operating system that she started a small business that provided consulting services for her "less-wired" peers. Her business card read, "I do Windows."

Bandwidth is different. It is not well understood, especially now that fiber optics is taking us from a modest to an almost infinite bandwidth, with nothing in between. Bandwidth is the capacity to move information down a given channel. Most people think of it by likening it to the diameter of a pipe or to the number of lanes on a highway.

These similes omit some of the more subtle and important differences among transmission media (copper, fiber, "air waves"). They ignore our ability to put more or fewer bits per second down the very same copper, fiber, or "air" pipe, depending on how we design (and modulate) the signal. Nonetheless, in general terms, we can characterize copper phone lines, fiber-optic connections, and the radio spectrum in some general ways that help us better understand the movement of our weightless bits.

Copper telephone wires, commonly called "twisted pair" (from the early days of their being braided like old lamp cords still found in old-fashioned, luxurious European hotels), are considered a low-bandwidth channel. Nonetheless, there is a $60 billion installed base of phone lines in America, capable of carrying up to 6 million bits per second with the appropriate modem (a word fabricated out of *modulator-demodulator,* the process of turning bits into waveforms and back again). A modem commonly runs at 9600 bits per second or 9600 baud. (Baud, named after Emile Baudot, the "Morse of Telex," is not technically the same as its per second [bps], but has come to be used interchangeably, as I have done in this text.)

Fancy modems can run at 38,400 baud (which is still more than a hundred times slower than the potential capacity of copper wire to most American homes). The way to think of twisted pair is like the turtle in the tale of the tortoise and the hare. It is slow, but not as slow as you may have been led to believe.

Think of the capacity of fiber as if it were infinite. We literally do not know how many bits per second we can send down a fiber. Recent research results indicate that we are close to being able to deliver 1,000 billion bits per second. This means that a fiber the size of a human hair can deliver every issue ever made of the *Wall Street Journal* in less than one second. Transmitting data at that speed, a fiber can deliver a million channels of television concurrently—roughly two hundred thousand times faster than twisted pair. That is a big jump. And mind you, I am talking of a single fiber, so if you want more, you just make more. It is, after all, just sand.

People assume that the transmission capacity of ether (popularly referred to as the "air waves") is infinite. It is air, after all, and there is so much of it everywhere. While I use the word *ether* throughout, it has only a historical meaning; once radio waves were discovered, ether was invoked as the mysterious substance that the waves traveled in; failure to actually find it helped lead to the discovery of photons. Stationary satellites orbit at 22,300 miles above the equator (which means about 34,000 billion cubic miles of ether contained in that outer envelope). That much ether must be able to carry a lot of bits without their bumping into one another. In one regard this is true when you think of the millions of remote-control units around the world that use wireless communication with TV sets and the like. Since these units have so little power, the few bits of data that pass from your hand to your TV do not change channels in the neighboring apartment or town. With cordless telephones, as many of us have heard, this is not quite the case.

As soon as we use the ether for higher-power telecommunications and broadcast, however, we have to be very careful that signals do not interfere with each other. We must be willing to live in predetermined parts of the spectrum, and we cannot use the ether piggishly. We must use it as efficiently as possible. Unlike fiber, we cannot manufacture any more of it. Nature did that once.

There are many ways to be efficient, like reusing parts of the spectrum by making a grid of transmission cells that allows people to use the same frequencies a few quadrants over or like moving into parts of the spectrum that were previously avoided (if only because those frequencies can fry innocent birds). But even with all the tricks and efficiencies, the bandwidth available in the ether is scarce by comparison with that provided by fiber and our endless ability to manufacture and lay more and more of it. For this reason, I proposed a trading of places between the wired and wireless information of today.

When Senator Bob Kerrey of Nebraska was campaigning for president, he spent a couple of hours at the Media Lab. When I met him, the first thing he said was, "The Negroponte Switch." That idea, which I first discussed and illustrated in a Northern Telecom meeting at which George Gilder and I were speakers, simply says that the information currently coming through the ground (read, wires) will come in the future through the ether, and the reverse. Namely, what is in the air will go into the ground and what is in the ground will go into the air. I called it "trading places." Gilder called it the "Negroponte Switch." The name stuck.

The reason I consider this trading of places self-evident is that bandwidth in the ground is infinite and in the ether it is not. We have one ether and an unlimited number of fibers. While we can be cleverer and cleverer with how we use the ether, in the end we ought to save all the spectrum we have for communication with things that move, which cannot be tethered, like a plane, boat, car, briefcase, or wristwatch.

FIBER: NATURE'S WAY

When the Berlin Wall came down six years ago, the *Deutsche Bundespost* bemoaned the timing as being five to seven years too soon. It was too early to build an all-fiber telephone plant throughout East Germany, as the prices were not yet low enough.

Today, fiber is cheaper than copper, including the cost of the electronics at each end. If you come across a condition where this is not true, wait a few months, as the prices of fiber connectors, switches, and transducers are plummeting. With the exception of communications lines a few feet or yards long or the presence of unskilled installers, there is no reason to use copper in telecommunications today (especially if you include the maintenance cost of copper). The Chinese are using fiber for a totally different reason—the villagers dig up the copper to sell on the black market.

The only real advantage of copper is the ability to deliver power. This is a touchy subject with telephone companies. Telephone companies are very proud of the fact that during a hurricane you may lose electric power, but your telephone is still likely to work. If your telephone used fiber instead of copper, it would need to get power from your local electric company and thus be vulnerable to blackouts. Even though you could have battery backup, that is an awkward solution because it needs special maintenance and attention. For this reason, you are likely to see the emergence of copper-shielded fiber or fiber-shielded copper. But from the perspective of bits, the entire wired planet will eventually be fiber.

One way to look at the transition from copper to fiber is to observe that American telephone companies replace roughly 5 percent of their plant each year, and they replace copper with fiber for maintenance and other reasons. While these upgrades are not evenly distributed, it is interesting to note that, at this rate, in just about twenty years the whole country could be fiber. The point is that we can quickly evolve a very broadband national infrastructure whether or not we need or know how to use that bandwidth. At the very least, a fiber plant will result in better-quality and more reliable "plain old telephone service" (which the insiders call POTS).

It has taken more than a decade to correct the mistake that Judge Harold Greene made in 1983 when he barred the Regional Bell Operating Companies from entering the information and entertainment industries. A major step forward was made on October 20, 1994, when the FCC approved so-called "video dialtone."

Ironically, the successful RBOC lobbyists used a gratuitous but effective argument to justify moving into the information and entertainment businesses. The telephone companies claimed that POTS was not enough and that unless they were allowed to become broader information providers, they could not justify the enormous cost of a new infrastructure (read, fiber).

Wait a minute. The phone companies have always been information providers. In fact, most of the RBOCs make their biggest profits from the Yellow Pages. But somehow, if they made this information into atoms and hurled them over the transom of your front door, that was okay. But if they kept this information as bits and delivered it to you electronically, that was illegal. Judge Greene saw it that way.

For this reason, lobbyists had to argue that the phone companies needed to be in the electronic information delivery business in order to justify the installation cost of a regional fiber plant. Without new revenue sources, the argument goes, there would be insufficient incentive to make this large investment. The argument worked, and the telephone companies are now marching into the information and entertainment businesses and laying fiber a little faster than before.

I think the result is good. It will benefit the consumer, but the argument was gratuitous. Phone companies may now be caught up in believing their own specious arguments against specious laws. We don't need these huge bandwidths to deliver most information and entertainment services. In fact, more modest bandwidth in the 1.2 to 6.0 million bps is well suited for most existing multimedia. We have not even started to understand or to tap the creative potential of 1.2 to 6.0 million bps. While lawyers and executives spent ten years pressing Judge Greene, they forgot to look at the enormous infrastructure already in place: twisted pair.

Few people realize how good copper phone lines are. A technique called ADSL (asymmetrical digital subscriber loop) is a means of sending large amounts of data down relatively short copper lines. ADSL-1 can provide 1.544 million bps into, and 64,000 bps out of, 75 percent of American and 80 percent of Canadian homes. ADSL-2 runs faster than 3 million bps, and ADSL-3 above 6 million bps. ADSL-1 is fine for VHS-quality video.

This is not a long-term solution for delivering multimedia to the home, but it is curious that it is being so widely ignored.

The explanation is said to be the high cost per subscriber. But that cost is a result of artificially low volumes. And even if the cost is temporarily high, like a thousand dollars per home, it is an incremental expense, much of which can be taken per household, as people subscribe. In addition, many Americans would be willing to pay some or all of the thousand dollars over the space of three or four years if the services were interesting, thereby sharing in the start-up cost. So while fiber is the future for sure, there is a lot we can be doing and learning with our existing copper plant today.

Many people are ignoring the copper stepping-stone. They are buying into the wholesale and immediate need for, and provision of, fiber for limitless bandwidth to maintain a major competitive edge, without recognizing that Mother Nature and commercial interests, more than regulatory incentives, will make fiber happen naturally. Like dogs in heat, broadband pundits are sniffing all the political opportunities for high-bandwidth networks as if doing so were a national imperative or civil right. In fact, unlimited bandwidth can have the paradoxical and negative effect of swamping people with too many bits and of allowing machines at the periphery to be needlessly dumb. Unlimited bandwidth is hardly wrong or bad to have, but like free sex, it is not necessarily good either. Do we really want or need all those bits?

LESS IS MORE

This expression of architect Mies van der Rohe finds its way into more and more lessons I have learned about the amount of information that needs to be transmitted and the means through which it is to be experienced. This is true of almost any new medium in the hands of a beginner. Beginners don't understand "less is more."

Take as an example the home video camera. When you first get and use a camcorder, you are very likely to pan and zoom a great deal, exercising simultaneously all the degrees of freedom you just discovered. The result is a manic, jerky home video, usually embarrassing to show (even the family winces at and is bored by the endless pans and zooms). Later, with the benefit of time, you calm down and use these new degrees of freedom more adroitly and sparingly.

Too much freedom has also had ill effects on our hard-copy output from laser printers. The ability to change font style and size is a temptation that pollutes many present-day university and business documents, which insensitively mix serif and sans serif type of all kinds and sizes: normal, bold, and italic, with and without shadows. It takes some deeper understanding of typography to realize that sticking with a single typeface is usually more appropriate, and to change size very sparingly. Less can be more.

Bandwidth is similar. There is a growing and ill-advised dogma that says we should use high bandwidth just because we have it. There really are some natural laws of bandwidth that suggest that squirting more bits at somebody is no more sensible or logical than turning up a radio's volume to get more information.

For example, 1.2 million bps in 1995 is a threshold for what has come to be called VHS-quality video. Go ahead and double or triple it, if you want better-quality television. It's hard to come up with a use for more than 6 million bps per person to deliver very new and imaginative services, if we have them.

New information and entertainment services are not waiting on fiber to the home; they are waiting on imagination.

COMPRESSING 100,000 BITS INTO 1

The relationship between bandwidth and computing is subtle. The trade-off between bandwidth and computing is evident today in video telephones and more expensive video conferencing systems. Because you compute at both ends of the line, you can ship fewer bits back and forth. By spending some money processing the digital video at each end, compressing and decompressing it, you use less channel capacity and save money in the transmission.

Digital video in general is an example of data compression without any attention to information content. People use the same coding techniques for an NFL game, a Ted Koppel–style interview, and a James Bond chase. Without being a computer scientist, you can probably guess that each of those programs would lend itself to very different approaches to data compression. As soon as content is considered, one can compress data very differently. Consider the following example taken from human-to-human discourse.

Imagine six animated people having dinner around a table; they are deeply engrossed in a common discussion about, say, a person not there. During one moment of this discourse about Mr. X, I look across the table at my wife and wink. After dinner, you come up to me and say, "Nicholas, I saw you wink at Elaine. What did you tell her?"

I explain to you that we had dinner with Mr. X two nights before, at which time he explained that, contrary to ———— he was in fact ————, even though people thought ————, but what he really decided was ———— etc. Namely, 100,000 bits (or so) later, I am able to tell you what I communicated to my wife with 1 bit (I ask your forbearance with my assumption that a wink is 1 bit through the ether).

What is happening in this example is that the transmitter (me) and the receiver (Elaine) hold a common body of knowledge, and thus communication between us can be in shorthand. In this example, I fire a certain bit through the ether and it expands in her head, triggering much more information. When you ask me what I said, I am forced to deliver to you all 100,000 bits. I lose the 100,000-to-1 data compression.

There is a story of a couple who knew hundreds of dirty jokes so well that they would merely recite numbers to each other. The few digits would call up an entire story and send one or the other into uncontrollable laughter. More prosaic use of this method in computer data compression is to number commonly used long words and send those few bits, instead of the entire string of letters. We are likely to see more and more such techniques when we trade bandwidth against shared knowledge. The condensation of information not only saves on the cost of shipping bits but saves our time as well.

THE ECONOMY OF SALES

Under today's methods of charging for telephone calls, I would pay a hundred thousand times more to send my story about

Mr. X to you than I would to send it to Elaine. The telecommunications companies have nothing to gain from shipping fewer bits back and forth. The current economic model of telephony is based on charging per second or per bit, irrespective of what the bit is.

The real question in understanding the economics of bandwidth is, Are some bits worth more than others? The answer is clearly yes. Yet a more complex question is, Should the value of a bit vary not only in accordance with its essential character (i.e., a movie bit, a conversation bit, or a pacemaker bit) but also in accordance with who is using it? or when? or how?

Most people, including those at *National Geographic,* would agree that a six-year-old child who uses their picture archive for homework should have access to those bits for free or almost for free. By contrast, if I were to use it for a paper or a business plan, I should pay a fair price and maybe even a tad extra to subsidize the six-year-old. Now the bits not only have a different value, but that value varies in accordance with who is using them and how. There are suddenly welfare bits, minority bits, and handicapped bits. Congress will have to be very creative in working out a framework for an equitable system.

Differential pricing of bits is not new. I have an account with Dow Jones that I use to log into the stock market. My account is embargoed for fifteen minutes. If I want up-to-date quotes like my eighty-six-year-old stockbroker uncle has, I have to pay a considerable premium to Dow Jones or to my uncle. This is the modern equivalent of the difference in cost between airmail and surface mail, bits that arrive by plane versus those that arrive by train.

In the case of real-time information, bandwidth requirements are dictated by the medium of discourse. If I am having a telephone conversation with you, it is meaningless to be able to get my voice to you faster than I can speak. Getting it to you slower or having a delay, of course, is intolerable. Even the quarter-second delay in a satellite telephone connection unsettles most people.

But if I record a message on tape, wish to transmit it to you, and am paying for that call by the minute, I would certainly want as many bits as possible to travel per second. This feeling is common to modem users who log in across the country to suck data into their laptops or squirt data out of them. Just a few years ago 2400 baud was considered very good. Today 38,400 bps is becoming common and results in a 94 percent reduction in phone charges.

Fortunately for the telephone companies, more than 50 percent of the telephonic traffic across the Pacific and 30 percent across the Atlantic is fax data running at 9600 bps, instead of 64,000 bps, which is also available.

STARS AND LOOPS

It's not just the bandwidth of the channels that matters, but their configuration as well. In simple terms, the phone system is a "star" network, which has lines radiating out from a point, like the avenues in Washington or the boulevards in Paris. There is literally a "home run" from your house to the nearest local telephone switch. If you wanted to, you could follow that twisted pair all the way back to your telephone company's local switching plant.

Cable television, by contrast, was born as a "loop," like a string of Christmas tree lights, that passes from home to home. These respective networks, stars and loops, assumed their shapes very naturally from the narrow bandwidth of twisted pair and the broader bandwidth of coaxial cable. In the first case, each home is served by a dedicated low-bandwidth line. In the second, a large number of households share a common broadband service.

The architecture of stars and loops is also driven by the nature of the content. In the case of the telephone network, each conversation is different, and the bits going to one home have absolutely no bearing on the others (minus one, perhaps). It is a vast-point-to-vast-point system by the nature of its operations. In the case of television, neighbors share the programming content, and it made all the sense in the world to have the Christmas-tree-light approach, a point-to-multipoint system. The conventional wisdom of cable operators has been largely to replicate terrestrial broadcasting as we know it, moving television from the ether onto their wires.

But conventional wisdom is just that: conventional. The future of television program delivery is changing radically, and you will not be satisfied either with the selection offered to your neighbor or by the need to view anything at a specific time. For this reason, cable companies are thinking more and more like telephone companies, with lots of switching and many home runs. In fact, twenty-five years from now, there may be no difference between cable and telephone, not only in the corporate sense but in terms of network architecture as well.

Eventually, most wiring will be stars. Loops will be deployed in very local areas or in the form of wireless broadcast, where the distribution medium by definition passes all homes at once. GM Hughes Electronics is fond of calling its satellite-based DirecTV system a "bent pipe," and will tell you how its direct-broadcast satellite television system is a cable system that passes every home in the United States. This is indeed true. At this very minute, if you are reading this in the United States, 1 billion bps are raining on you from that Hughes bird, unless you are under a lead umbrella.

PACKAGING BITS

Many people who have taken a small step toward being digital think of bandwidth like plumbing. Thinking of bits like atoms leads to big pipes and little pipes, faucets and hydrants. A commonly used comparison is that using fiber is like drinking from a fire hose. The analogy is constructive but misleading. Water flows or doesn't flow. You can regulate how much comes out of a garden hose by closing the nozzle. But even as the flow from a fire hose slows to a trickle, the water atoms are moving as a group.

Bits are different. A ski lift may be a better analogy. The lift is moving at a constant speed, while more or fewer people get on and off. Similarly, you put a number of bits into a packet and then drop that packet in a pipe capable of delivering it at a speed of millions of bits per second. Now, if I drop a packet of 10 bits every second into a fast-moving pipe, my effective bandwidth is 10 bps, not the speed of the pipe.

While this sounds wasteful, it is in fact a clever notion, because other people are dropping packets into the same pipe—

the basis of such systems as the Internet and ATM (asynchronous transfer mode, the way all telephone networks will work in the near future). Instead of tying up an entire telephone line, as you now do for voice, packets are put into the queue with names and addresses attached to them, so they know when and where to get off this ski lift. You pay for packets, not minutes.

Another way of thinking of the same packetizing of bandwidth: the best way to use a billion bits per second is to use a thousand bits in a millionth of a second, a million bits in a thousandth of a second, or the like. In the case of television, for example, think of receiving one hour of video in a few seconds, versus the faucet approach.

Instead of delivering a thousand television programs to everybody, it may be better to deliver one program to each person in one-thousandth of the real time. This will totally change how we think of broadcast media. The broadcast of most bits will have absolutely nothing to do with the rate at which we consume them as humans.

BITCASTING   3

WHAT'S WRONG WITH THIS PICTURE?

When you watch television, do you complain about picture resolution, the shape of the screen, or the quality of motion? Probably not. If you complain, it is surely about programming. Or, as Bruce Springsteen says, "fifty-seven channels and nothin' on." Yet almost all the research directed at the advance of television is aimed precisely at refining the display as opposed to the artistry of content.

In 1972 a few visionary Japanese asked themselves what the next evolutionary step in television would be. They reached the conclusion that it would be higher resolution, postulating that the move from black-and-white to color would be followed by filmic-quality TV or so-called high-definition television (HDTV). In an analog world this was a logical way to scale up television, and it is what the Japanese did for the next fourteen years, calling it Hi-Vision.

In 1986 Europe was alarmed by the prospect of Japanese dominance of a new generation of television. Worse, the United States embraced Hi-Vision and lobbied with the Japanese for it to become a world standard. Many present-day proponents of American HDTV and most neonationalists conveniently forget this misjudgment of backing a Japanese analog system. As a purely protectionist measure, the Europeans voted Hi-Vision down, doing us all a great favor, albeit for the wrong reason. They then proceeded to develop their own analog HDTV system—HD-MAC—which in my view was slightly worse than Hi-Vision.

More recently, the United States, like a sleeping giant, awoke and attacked the HDTV problem with the same analog abandon as the rest of the world, becoming third in line to address the future of television as nothing more than a problem of picture quality and, worse, to tackle that problem with old-fashioned analog techniques. Everybody assumed that increased image quality was the relevant course to pursue. Unfortunately, this is not the case.

There is no proof to support the premise that consumers prefer better picture quality rather than better content. This is particularly true given that the solutions so far proposed for HDTV may not even result in enough noticeable image improvement, compared with the studio-quality television available today (which you probably have never seen and might not guess how good it is). HDTV at the current level of HD is just silly.

THE LAST SHALL BE FIRST

In 1990 we were presented with the likely outcome of Japan, Europe, and the United States going in totally different directions regarding advanced television. Japan had by then invested eighteen years of money and effort on HDTV. During that time, the Europeans had seen the computer industry slip out of their grasp and were determined not to have this happen with television. And the United States, which had virtually no television industry at all, saw HDTV as the great opportunity for re-entry into consumer electronics (which shortsighted companies like Westinghouse, RCA, and Ampex had earlier just given away).

When America took up the challenge of improving television technology, digital compression was at too early a stage of development to be the obvious course of action. Also, the protagonists, TV equipment manufacturers, were just the wrong players. Unlike young digital companies such as Apple and Sun Microsystems, television technology companies were old-age homes for analog thought. To them, television was about pictures, not about bits.

But shortly after the American awakening, in 1991, almost overnight, everybody became a proponent of digital television, following the lead of General Instrument Corporation. Literally within less than six months, each American proposal for HDTV was changed from analog to digital. There was just enough evidence that digital signal processing would be cost-effective, which is something Europe argued against until February 1993.

In September 1991, I addressed many members of President François Mitterrand's cabinet over lunch. Perhaps because I speak French only as a second language, I could not convince them that I was not trying to get them to give up their "lead," as they called it, but to get rid of "the anchor around their necks," as I called it.

When I met with the Japanese Prime Minister Kiichi Miyazawa in 1992, he was startled to learn that Hi-Vision was obsolete. Margaret Thatcher, however, did listen to me. Finally, in late 1992 the tide was turned by John Major's bold move to veto an ECU 600 million ($800 million) subsidy of HDTV programming. The European Union (then called the European Community) finally decided in early 1993 to abandon analog HDTV in favor of a digital future.

The Japanese know full well that digital TV is the future. When Akimasa Egawa, the hapless director general of Posts and Telecommunications' Broadcasting Administration Bureau, suggested in February 1994 that Japan join the digital world, Japanese industrial leaders cried foul the next day and forced him to eat his own words. Japan had spent so much public money on HDTV, it was not about to cut its losses so publicly.

I recall vividly at the time a televised panel of the presidents of the giant consumer electronics companies swearing they were fully behind good old analog Hi-Vision, implying that the deputy minister was off his rocker. I had to bite my digital tongue, because I knew each of them personally, had heard them say the opposite, and had seen their respective digital TV efforts. Saving face, I fear, is to have two of them.

RIGHT TECHNOLOGY, WRONG PROBLEMS

The good news is that in the United States we are applying the right technology, digital, to the future of television. The bad news is that we are still mindlessly addressing the wrong problems, those of image quality—resolution, frame rate, and the shape of the screen (the so-called aspect ratio). Worse, we are trying to decide once and for all on very specific numbers for each and to legislate these variables as constants. The great gift of the digital world is that you don't have to do this.

Even the analog world is getting less stubborn. Anybody who has traveled in Europe remembers the terrible problem of transformers, adapting 220 volts to our 110-volt appliances. The story goes that Don Estridge, the IBM executive who fathered the IBM PC, was in the parking lot of the company's Boca Raton, Florida, facilities one day and demanded that the PC not care whether the power was 110 or 220. This seemingly outlandish command was implemented shortly thereafter, and today almost any personal computer can be plugged into a wide range of power sources. One way to think of this is that Estridge's request was met by putting that intelligence in the machine (let the plug worry about what the human previously worried about). Now for a message to TV makers.

More and more we will see systems that have the ability to adapt, not just to 110 or 220 volts, 60 Hz and 50 Hz, but to the number of scan lines, the frame rate, and the aspect ratio. The equivalent already happens with modems, which do a great deal of handshaking with each other to settle on the best possible communications protocols. By extension, this also happens in e-mail, where systems use a variety of protocols for passing messages between different machines, with greater or lesser success—but almost never with none at all.

Being digital is the license to grow. At the onset, you need not dot every i and cross every t. You can build hooks for future expansion and develop protocols so that bit streams can tell each other about themselves. Digital TV pundits have ignored this property. Not only are they working on the wrong problem, high definition, they are taking all the other variables and treating them like the 110 volts of hair dryers.

Arguments about "interlace" are a perfect example. TV is 30 frames per second. Each frame is composed of two so-called fields, each with half the scan lines (the odds and the evens). Therefore, a video frame is composed of two fields offset by one scan line and offset in time by one-sixtieth of a second. When you watch television, you are seeing 60 fields per second (so motion is smooth) "interlaced" together, but each field has only half of the picture. The result is that you perceive good-quality motion and you see stationary objects very clearly with only half the bandwidth—a great idea for the television broadcast when it was analog and when bandwidth is at a premium.

The dilemma comes from computer displays, for which interlace is meaningless and harmful to moving images. Computer displays need to be more accurate (greater resolution and viewable from a much closer vantage point), and motion plays a very different role on our closely watched computer screens. Suffice it to say, interlace has no future whatsoever with computers and is rightfully anathema to any computer engineer.

But interlace will die a natural death. To pass a law against it would be about as reasonable as a blue law. The digital world is far more resilient than the analog domain in that signals can carry all sorts of additional information about themselves. Computers can process and postprocess signals, add and subtract interlace, change the frame rate, and modify the aspect ratio to match the rectangular form factor of a particular signal with the shape of a particular display. For this reason, we are better off making as little as possible part of some arbitrarily fixed standard, if only because what sounds logical today will prove to be nonsense tomorrow.

AS SCALABLE AS THE U.S. CONSTITUTION

The digital world is intrinsically scalable. It can grow and change in a more continuous and organic way than former analog systems. When you buy a new TV set, you throw away one and adopt a totally new one. By contrast, if you have a computer, you are accustomed to adding features, hardware and software, instead of exchanging everything for the tiniest upgrade. In fact, the word *upgrade* itself has a digital tone to it. We are more and more accustomed to scaling computer systems up, getting a better display, installing enhanced sound, and fully expecting our software to work better, versus not at all. Why isn't TV like that?

It will be. Today we are stuck with three analog TV standards. In the United States and Japan we use NTSC (which stands for National Television Systems Committee, although Europeans will tell you it means "Never the Same Color"). PAL (Phase Alternating Line) dominates Europe and is trailed in France by SECAM (SEquential Couleur Avec Memoire); Americans have been known to say it really means "Something Essentially Contrary to America." The rest of the world follows willy-nilly, using one of the three in its pure or impure form, with almost as much logic as the national choice of a second language.

Being digital is the option to be independent of confining standards. If your TV does not speak a particular dialect, you may have to visit your local computer store and buy a digital decoder, just like you buy software for your PC today.

If resolution is an important variable, then surely the solution is to build a scalable system, not one glued to the number of scan lines we can display easily today. When you hear people talk about 1,125 or 1,250 scan lines, there is nothing magical about those numbers. They just happen to be close to the maximum we can display with a cathode ray tube (CRT) today. In fact, the way TV engineers have considered scan lines in the past is no longer operative.

In the old days, as TV sets got bigger, the viewer moved farther away from it, ultimately to the proverbial couch. On average, the number of scan lines per millimeter that hit the viewer's pupil was more or less constant.

Then, around 1980, there was a sudden change and people were brought off the couch to the desktop, for an eighteen-inch viewing experience. This change turns scan line thinking upside down, because we can no longer think of scan lines per picture (like we always have with TV sets), but scan lines per inch, as we do with paper or modern computer displays (for which Xerox Corporation's Palo Alto Research Center, PARC, gets the credit of being the first to think in terms of lines per inch). A bigger display needs more lines. Eventually, when we can tile together flat-panel displays, we will have the ability to present images with ten thousand lines of resolution. To limit our thinking to roughly a thousand today is shortsighted.

The way to achieve massively high resolution tomorrow is to make the system scalable today, which is exactly what none of the current proponents of digital TV systems are proposing at the moment. Odd.

TV AS A TOLLBOOTH

All makers of computer hardware and software are courting the cable industry, which is not surprising when we consider that ESPN has more than 60 million subscribers. Microsoft, Silicon Graphics, Intel, IBM, Apple, DEC, and Hewlett-Packard have all entered major agreements with the cable industry.

The object of this ferment is the set-top box, currently little more than a tuner but destined to be much more. At the rate things have been going, we will soon have as many types of set-top boxes as we now have infrared remote-control units (one for cable, one for satellite, one for twisted pair, one for every UFV transmission, etc.). Such a smorgasbord of incompatible set-top boxes is a horrible thought.

The interest in this box stems from its potential function as, among other things, a gateway through which the "provider" of that box and its interface can become a gatekeeper of sorts, charging onerous fees for information as it passes through this tollgate and into your home. While this sounds like a dandy business, it is unclear if it's in the public's best interest. Worse, a set-top box itself is technically shortsighted and the wrong focus. We should broaden our vision and set our sights instead on general-purpose and less proprietary computer designs.

The word *box,* as in "set-top box," carries all the wrong connotations, but here's the theory. Our insatiable appetite for bandwidth puts cable television currently in the lead position as the broadband provider of information and entertainment services. Cable services today include set-top boxes because only a fraction of TV receivers are cable-ready. Given the existence and acceptance of this box, the idea is simply to aggrandize it with additional functions.

What's wrong with this plan? It's simple. Even the most conservative broadcast engineers agree that the difference between a television and computer will eventually be limited to peripherals and to the room of the house in which it is found. Nevertheless, this vision is being compromised by the cable industry's monopolistic impulses and by an incremental improvement of a box to control 1,000 programs, 999 of which at any one moment you are not (by definition) watching. In the lucrative sport of making digital television, the computer has so far been seriously "out-boxed" in the first round.

But its comeback will be triumphant.

THE TV AS COMPUTER

I am fond of asking people if they remember Tracy Kidder's book *The Soul of a New Machine.* I then ask somebody who has read it if he can remember the name of the computer company in question. I have yet to meet somebody who can. Data General (that's the one), Wang, and Prime, once high-flying, high-growth companies, had a total disregard for open systems. I remember sitting around boardroom tables where people argued that proprietary systems would be a great competitive advantage. If you could make a system that was both popular and unique, you would lock out competition. Sounds logical, but it is totally wrong, and that is why Prime no longer exists, and the other two, as well as many others, are shadows of their former selves. This is also why Apple is changing its strategy today.

"Open systems" is a vital concept, one that exercises the entrepreneurial part of our economy and challenges both proprietary systems and broadly mandated monopolies. And it is winning. In an open system we compete with our imagination, not with a lock and key. The result is not only a large number of successful companies but a wider variety of choice for the consumer and an ever more nimble commercial sector, one capable of rapid change and growth. A truly open system is in the public domain and thoroughly available as a foundation on which everybody can build.

The growth of personal computers is happening so rapidly that the future open-architecture television is the PC, period. The set-top box will be a credit-card-size insert that turns your PC into an electronic gateway for cable, telephone, or satellite. In other words, there is no TV-set industry in the future. It is nothing more or less than a computer industry: displays filled with tons of memory and lots of processing power. Some of those computer products may be ones with which you are more likely to have a ten-foot, rather than an eighteen-inch, experience, more often in a group than as an individual. But any way you look at it, it's still a computer.

The reason is that computers are becoming more and more video enabled, equipped to process and display video as a data type. For teleconferencing, multimedia publications, and a host of simulation applications, video is becoming part of all, not just many, computers. This is happening so fast that the snail's pace of television development, albeit digital, will be eclipsed by the personal computer.

The rhythm of HDTV development, for example, has been synchronized with the Olympic Games, in part to get international exposure and in part to be seen in one of its more credible lights: spectacle sporting events. You might not actually be able to see the hockey puck on normal television. Thus Japan used the Seoul Summer Olympics in 1988 to launch Hi-Vision, and the Europeans introduced HD-MAC at the Albertville Winter Olympics in 1992 (only to discontinue it less than one year later).

The American HDTV players have proposed the summer of 1996, during the Atlanta Olympics, as the time to feature their new digital, closed-architecture HDTV system. By then it will be far too late, and HDTV will be stillborn. By then, nobody will care, and as many as 20 million Americans could be watching NBC in a window in the upper-right-hand corner of their personal computer screens. Intel and CNN jointly announced such a service in October 1994.

THE BIT RADIATION BUSINESS

The key to the future of television is to stop thinking about television as television. TV benefits most from thinking of it in terms of bits. Motion pictures, too, are just a special case of data broadcast. Bits are bits.

The six o'clock news not only can be delivered when you want it, but it also can be edited for you and randomly accessed by you. If you want an old Humphrey Bogart movie at 8:17 p.m., the telephone company can provide it over its twisted pair. Eventually, when you watch a baseball game, you will be able to do so from any seat in the stadium or, for that matter, from the perspective of the baseball. These are the kinds of changes that come from being digital, as opposed to watching "Seinfeld" at twice today's resolution.

When television is digital, it will have many new bits— the ones that tell you about the others. These bits may be simple headers that tell you about resolution, scan rate, and aspect ratio, so that your TV can process and display the signal to its fullest capacity. These bits may be the decoding algorithm that lets you see some strange or encrypted signal when combined with the bar code from a box of corn flakes. The bits may be from one of a dozen sound tracks that enable you to watch a foreign movie in your own language. The bits may be the control data for a knob that allows you to change X-rated to R-rated to PG-rated material (or the reverse). Today's TV set lets you control brightness, volume, and channel. Tomorrow's will allow you to vary sex, violence, and political leaning.

Most television programs, with the exception of sporting events and election results, need not be in real time, which is crucial to digital television and largely ignored. This means that most TV is really like downloading to a computer. The bits are transferred at a rate that has no bearing on how they will be viewed. More important, once in the machine, there is no need to view them in the order they were sent. All of a sudden TV becomes a random access medium, more like a book or newspaper, browsable and changeable, no longer dependent on time or day, or the time required for delivery.

Once we stop thinking of TV's future as only high definition and begin to build it in its most general form, bit radiation, TV becomes a totally different medium. We will then start to witness many creative and engaging new applications on the information superhighway. That is, unless we are stopped by the Bit Police.

THE BIT POLICE

4

THE LICENSE TO RADIATE BITS

Thereare five paths for information and entertainment to get into the home: satellite, terrestrial broadcast, cable, telephone, and packaged media (all those atoms, like cassettes, CDs, and print). The Federal Communications Commission, the FCC, serves the general public by regulating some of these paths and some of the information content that flows over them. Its job is a difficult one because the FCC is often at the thorny edges between protection and freedom, between public and private, between competition and broadly mandated monopolies.

One area of major FCC concern is the spectrum used for wireless communications. The spectrum is considered something that belongs to everybody and should be used fairly, competitively, without interference, and with all the incentives to be enriching for the American people. This makes perfect sense, because without such oversight, television signals, for example, might collide with cellular telephone, or radio might interfere with marine VHF. The highway in the sky really does need some air traffic control.

Recently, some parts of the spectrum have been auctioned at very high prices for cellular telephony and interactive video. Other parts of the spectrum are given away free, in that they are said to serve the public interest. This is the case of advertiser-supported television, which is "free" to the viewer. In fact, you are paying for it when you buy a box of Tide or any other advertised product.

The FCC has proposed to give existing television broadcasters an additional "lane," 6 MHz (megahertz), of free spectrum for HDTV, on the condition that their currently used spectrum, also 6 MHz, is returned within fifteen years. Namely, for that fifteen-year period, existing broadcasters would have 12 MHz. The idea, subject to change, is to provide a transition period for current television to evolve into future television. The concept made perfectly good sense six years ago, when it was conceived as the path to go from one analog world to another. But suddenly HDTV is digital. We now know how to deliver 20 million bps in a 6 MHz channel, and all the rules may suddenly change, in some cases in a quite unanticipated manner.

Imagine that you own a TV station and the FCC just gave you a license to broadcast 20 million bps. You have just been given permission to become a local epicenter in the bit radiation business. This license was meant for television, but what would you really do with it?

Be honest. The very last thing you would do is broadcast HDTV, because the programs are scarce and the receivers few. With a little cunning, you'd probably realize that you could broadcast four channels of digital, studio-quality standard NTSC television (at 5 million bps each), thereby increasing your potential audience share and advertising revenue. Upon further reflection, you might instead decide to transmit three TV channels, using 15 million bps, and devote the remaining 5 million bps to two digital radio signals, a stock-data broadcast system, and a paging service.

At night, when few people are watching TV, you might use most of your license to spew bits into the ether for delivery of personalized newspapers to be printed in people's homes. Or, on Saturday, you might decide that resolution counts (say, for a football game) and devote 15 million of your 20 million bits to a high-definition transmission. Literally, you could be your own FCC for that 6 MHz or those 20 million bits, allocating them as and when you see fit.

This is not what the FCC originally had in mind when it recommended allocation of the new HDTV spectrum to existing broadcasters for transition purposes. Groups hankering to get into the bit radiation business will scream bloody murder when they realize that current TV stations just had their spectrum doubled and their broadcast capacity increased by 400 percent, at no cost, for the next fifteen years!

Does that mean we should send in the Bit Police to make sure that this new spectrum and all of its 20 million bps are used only for HDTV? I hope not.

BITS OF CHANGE

In analog days, the spectrum allocation part of the FCC's job was much easier. It could point to different parts of the spectrum and say: this is television, that is radio, this is cellular telephony, etc. Each chunk of spectrum was a specific communications or broadcast medium with its own transmission characteristics and anomalies, and with a very specific purpose in mind. But in a digital world, these differences blur or, in some case, vanish: they are all bits. They may be radio bits, TV bits, or marine communication bits, but they are bits nonetheless, subject to the same commingling and multi-use that define multimedia.

What will happen to broadcast television over the next five years is so phenomenal that it's difficult to comprehend. It is hard to imagine that the FCC can or will regulate the bits by, for example, demanding quotas of bits be used for HDTV, normal TV, radio, and so on. Surely the market is a much better regulator. You would not use all of your 20 million bps for radio if there were better revenue in TV or data. You would find yourself changing your own allocation depending on the day of the week, the time of the day, holidays, or special events. The flexibility is crucial, and the public will be served best by those who are quickest to respond and most imaginative in the use of their bits.

In the near future, broadcasters will assign bits to a particular medium (like TV or radio) at the point of transmission. This is usually what people mean when they talk about digital convergence or bit radiation. The transmitter tells the receiver, here come TV bits, here comes radio, or here come bits that represent the *Wall Street Journal*.

In the more distant future the bits will not be confined to any specific medium when they leave the transmitter.

Take the weather as an example. Instead of broadcasting the weatherman and his proverbial maps and charts, think of sending a computer model of the weather. These bits arrive in your computer-TV and then you, at the receiving end, implicitly or explicitly use local computing intelligence to transform them into a voice report, a printed map, or an animated cartoon with your favorite Disney character. The smart TV set will do this in whatever way you want, maybe even depending on your disposition and mood at the moment. In this example, the broadcaster does not even know what the bits will turn into: video, audio, or print. You decide that. The bits leave the station as bits to be used and transformed in a variety of different ways, personalized by a variety of different computer programs, and archived or not as you see fit.

That scenario truly is one of bitcasting and datacasting and beyond the kind of regulatory control we have today, which assumes the transmitter knows that a signal is TV, radio, or data.

Many readers may have assumed that my mention of Bit Police was synonymous with content censorship. Not so. The consumer will censor by telling the receiver what bits to select. The Bit Police, out of habit, will want to control the medium itself, which really makes no sense at all. The problem, strictly political, is that the proposed HDTV allocation looks like a handout. While the FCC had no intention of creating a windfall, special-interest groups will raise hell because the bandwidth rich are getting bandwidth richer.

I believe the FCC is too smart to want to be the Bit Police. Its mandate is to see advanced information and entertainment services proliferate in the public interest. There is simply no way to limit the freedom of bit radiation, any more than the Romans could stop Christianity, even though a few brave and early data broadcasters may be eaten by the Washington lions in the process.

CROSS-OWNERSHIP

Consider a modern newspaper. The text is prepared on a computer; stories are often shipped in by reporters as e-mail. The pictures are digitized and frequently transmitted by wire as well. And the page layout of a modern newspaper is done with computer-aided design systems, which prepare the data for transfer to film or direct engraving onto plates. This is to say that the entire conception and construction of the newspaper is digital, from beginning to end, until the very last step, when ink is squeezed onto dead trees. This is the step where bits becomes atoms.

Now imagine that the last step does not happen in a printing plant, but that the bits are delivered to you as bits. You may elect to print them at home for all the conveniences of hard copy (for which reusable paper is recommended, so we all don't need a large pile of blank newsprint). Or you may prefer to download them into your laptop, palmtop, or someday into your perfectly flexible, one-hundredth-of-an-inch-thick, full-color, massively high-resolution, large-format, waterproof display (which just happens to look exactly like a sheet of paper and smell like one, too, if that's what turns you on). While there are many ways to get you the bits, one is surely broadcast. The television broadcaster can send you newspaper bits.

Oops. Generally, cross-ownership rules say you cannot own a newspaper and a TV station in the same place. In the analog days, the easiest mechanism for preventing monopoly and for guaranteeing plurality and multiple voices was to limit an owner to a single medium in any one town or city. Media diversity meant content diversity. So, if you owned the newspaper, you could not own the TV station, and vice versa.

In 1987, Senators Ted Kennedy and Ernest Hollings added an eleventh-hour rider to a continuing budget resolution that prevented the FCC from extending temporary waivers of its cross-ownership regulation. This was targeted at Rupert Murdoch, who had purchased a newspaper in Boston while owning a UHF station there. The so-called laser-beam law directed at Murdoch was overturned by the courts a few months later, but the congressional ban against the FCC changing or repealing cross-ownership rules remains.

Should it really be unlawful to own a newspaper bit and a television bit in the same place? What if the newspaper bit is an elaboration of the TV bit in a complex, personalized multimedia information system? The consumer stands to benefit from having the bits commingle and the reporting be at various levels of depth and display quality. If current cross-ownership policies remain in existence, isn't the American citizen being deprived of the richest possible information environment? We are shortchanging ourselves grotesquely if we forbid certain bits to commingle with others.

Guaranteed plurality might require less legislation than one would expect, because the monolithic empires of mass media are dissolving into an array of cottage industries. As we go online and deliver more and more bits and fewer and fewer atoms, the leverage of owning printing plants will disappear. Even having a dedicated staff of reporters worldwide will lose some of its significance as talented free-lance writers discover an electronic venue directly into your home.

Media barons of today will be grasping to hold on to their centralized empires tomorrow. I am convinced that by the year 2005 Americans will spend more hours on the Internet (or whatever it is called) than watching network television. The combined forces of technology and human nature will ultimately take a stronger hand in plurality than any laws Congress can invent. But in case I'm wrong in the long term and for the transition period in the short term, the FCC had better find some imaginative scheme to replace industrial-age cross-ownership laws with incentives and guidelines for being digital.

BIT PROTECTION?

Copyright law is totally out of date. It is a Gutenberg artifact. Since it is a reactive process, it will probably have to break down completely before it is corrected.

Most people worry about copyright in terms of the ease of making copies. In the digital world, not only the ease is at issue, but also the fact that the digital copy is as perfect as the original and, with some fancy computing, even better. In the same way that bit strings can be error-corrected, a copy can be cleaned up, enhanced, and have noise removed. The copy is perfect. This is well known to the music industry and has been the cause of delaying several consumer electronics products, notably DAT (digital audiotape). This may be senseless, because illegal duplication seems to be rampant even when the copies are less than perfect. In some countries, as many as 95 percent of all videocassettes sold are pirated.

The management of and attitude toward copyrights today vary dramatically from medium to medium. Music enjoys considerable international attention, and the creative people who make melodies, lyrics, and sounds get reimbursed for years. The melody for "Happy Birthday" is in the public domain, but if you want to use the lyrics in the scene of a movie, you must pay Warner/Chappell a royalty. Not very logical, but nonetheless part of a complex system of protecting music composers and performers.

By contrast, a painter more or less kisses a painting good-bye upon its sale. Pay-per-view would be unthinkable. On the other hand, in some places it is still perfectly legal to chop up the painting and resell it in smaller pieces, or to replicate it as a carpet or beach towel without the artist's permission. In the United States, it was not until 1990 that we enacted a Visual Artists Rights Act to prevent that kind of mutilation. So even in the analog world the current system is not very long-standing or completely evenhanded.

In the digital world it is not just a matter of copying being easier and copies more faithful. We will see a new kind of fraud, which may not be fraud at all. When I read something on the Internet and, like a clipping from a newspaper, wish to send a copy of it to somebody else or to a mailing list of people, this seems harmless. But, with less than a dozen keystrokes, I could redeliver that material to literally thousands of people all over the planet (unlike a newspaper clipping). Clipping bits is very different from clipping atoms.

In the irrational economics of today's Internet, it costs exactly zero pennies to do the above. Nobody has a clear idea of who pays for what on the Internet, but it appears to be free to most users. Even if this changes in the future and some rational economic model is laid on top of the Internet, it may cost a penny or two to distribute a million bits to a million people. It certainly will not cost anything like postage or FedEx rates, which are based on moving atoms.

Furthermore, computer programs, not just people, will be reading material such as this book and making, for example, automatic summaries. Copyright law says that if you summarize material, that summary is your intellectual property. I doubt that lawmakers ever considered the idea of abstracting being done by an inanimate entity or robo-pirates.

Unlike patents, which in the United States reside in a totally different branch of government (Department of Commerce, hence executive) than copyrights (Library of Congress, hence legislative), copyrights protect the expression and form of the idea, as opposed to the idea itself. Fine.

What happens when we transmit bits that are in a real sense formless, like the weather data referred to earlier? I am hard-pressed to say whether a computer model of the weather is an expression of the weather. In fact, a complete and robust computer model of the weather is best described as a simulation of the weather and is as close to the "real thing" as can be imagined. Surely, the "real thing" is not an expression of itself, but is itself.

The expressions of the weather are: a voice "speaking" it with intonation, an animated diagram "showing" it with color and motion, or a mere printout "depicting" it as an illustrated and annotated map. These expressions are not in the data, but are embodiments of them made by a quasi (or really) intelligent machine. Furthermore, these different incarnations may reflect you and your expressive tastes, versus those of a local, national, or international weather forecaster. There is nothing to copyright at the transmitter.

Take the stock market. The minute-to-minute fluctuations of share prices can be assembled in a variety of different ways. The body of data, like the contents of the phone company's white pages, is not copyrightable. But an illustration of the performance of a stock or group of stocks is most definitely copyrightable. That kind of form will be increasingly given to data by the receiver, not the transmitter, and further complicates the problem of protection.

To what degree can the notion of formless data be extended to less prosaic material? To a news report (possible) or to a novel (harder to imagine)? When bits are bits, we have a whole new suite of questions, not just the old ones like piracy.

The medium is no longer the message.

COMMINGLED BITS 5

REPURPOSING THE MATERIAL GIRL

The fact that, in one year, a then thirty-four-year-old former Michigan cheerleader generated sales in excess of $1.2 billion did not go unnoticed by Time Warner, which signed Madonna to a $60 million "multimedia" contract in 1992. At the time, I was startled to see *multimedia* used to describe a collection of unrelated traditional print, record, and film productions. Since then, I see the word almost every day in the *Wall Street Journal,* often used as an adjective to mean anything from interactive to digital to broadband. One headline read, "Record Shops Yield to Multimedia Stores." It would seem that if you are an information and entertainment provider who does not plan to be in the multimedia business, you will soon be out of business. What is this all about?

It's both about new content and about looking at old content in different ways. It's about intrinsically interactive media, made possible by the digital lingua franca of bits. And it's about the decreasing costs, increasing power, and exploding presence of computers.

This technological pull is augmented by an aggressive push from media companies, which are selling and reselling as many of their old bits as possible, including Madonna's (which sell so well). This means not only reuse of music and film libraries but also the expanded use of audio and video, mixed with data, for as many purposes as possible, in multiple packages and through diverse channels. Companies are determined to repurpose their bits at a seemingly small marginal cost and at a likely large profit.

If thirty minutes of situation comedy costs CBS or Fox half a million dollars, it takes very little wisdom to conclude that your existing library of, say, ten thousand hours of film material might be reused profitably. If you were to value your old bits very conservatively at one-fiftieth the cost of the new ones, that makes your library worth $200 million. Not bad.

Repurposing goes hand in hand with the birth of any new medium. Film reused plays, radio resold performances, and TV recycled movies. So there is nothing unnatural about Hollywood's yearning to repurpose its video archives or to combine them with music and text. The problem is that indigenous multimedia material, native to this new medium, is hard to come by in these early days.

Information and entertainment services that really take advantage of and define new multimedia must evolve and need a gestation period long enough to accommodate both successes and failures. As a consequence, multimedia products today are like newborn children with good genes, but not yet sufficiently developed to have a recognizable character and strong physique. Most of today's multimedia applications are somewhat anemic, rarely more than one kind of opportunism or another. But we are learning fast.

From a historical perspective, the incubation period of a new medium can be quite long. It took many years for people to think of moving a movie camera, versus just letting the actors move in front of it. It took thirty-two years to think of adding sound. Sooner or later, dozens of new ideas emerged to give a totally new vocabulary to film and video. The same will happen with multimedia. Until we have a robust body of such concepts, we are bound to see considerable regurgitation of archival bits. This may be okay with *Bambi* bits, but not so interesting with those of *Terminator 2*.

Delivering child's fare as multimedia in the form of CD-ROMs (i.e., in atom form) works particularly well because a child is willing to look at or listen to the same story time and time again. I had one of the first Pioneer LaserDisc players at home in 1978. At that time, only one movie existed on laser disc: *Smokey and the Bandit*. My then eight-year-old son was fully prepared to look at this movie hundreds of times, to the point where he discovered impossible cuts (Jackie Gleason on one side of the car door in one frame and on the other in the next frame), which just escapes you at 30 frames per second. In a later release, *Jaws*, he was able to find the wiring of the shark by single framing, thus occupying himself for many hours.

During this period, *multimedia* meant trendy electronic nightclubs, with strobe lights and glitz. It carried the connotation of rock music plus light show. I was specifically asked to remove the word *multimedia* from a proposal to the Department of Defense. DOD staff were afraid I would get the notorious Golden Fleece Award from Senator William Proxmire, an annual prize given to the most gratuitously funded government projects, and all the negative publicity that came with it. (In December 1979 the Office of Education, so-called at the time, was less lucky when one of their researchers won the Fleece for spending $219,592 to develop a "curriculum package" to teach college students how to watch television.)

When we showed a fully colored and illustrated page of text on a computer screen, people gasped in astonishment when the illustration turned into a sound-synch movie at the touch of a finger. Some of today's best multimedia titles are high-production-value renditions of less well-made but seminal experiments of that period.

BIRTH OF MULTIMEDIA

Late at night on July 3, 1976, the Israelis launched an extraordinarily successful strike on the Entebbe, Uganda, airport, rescuing 103 hostages taken prisoner by pro-Palestinian guerrillas, who were given safe haven by dictator Idi Amin. By the time the one-hour operation ended, twenty to forty Ugandan troops were killed and all seven hijackers were dead. Only one Israeli soldier and three hostages also lost their lives. This impressed the American military so much that the Advanced Research Projects Agency, ARPA, was asked to investigate electronic ways in which American commandos could get the kind of training the Israelis had had to succeed at Entebbe.

What the Israelis had done was build in the desert a physical model, to scale, of the Entebbe Airport (which was easy for them to do because Israeli engineers had designed the airport when the two nations were on friendly terms). The commandos then practiced landings and takeoffs, as well as simulated assaults on this accurate mock-up. By the time they arrived in Uganda for the "real thing," they had an extraordinarily keen spatial and experiential sense of the place, allowing them to perform like natives. What a simple and terrific idea.

However, the idea as a physical embodiment was not extensible, in that we just could not build replicas of every potential hostage situation or terrorist targets like airports and embassies. We needed to do this with computers. Once again, we had to use bits, not atoms. But computer graphics alone, like that used in flight simulators, was inadequate. Whatever system was developed would need the full photorealism of a Hollywood stage set to convey a real sense of place and a feel for the surrounding environment.

My colleagues and I proposed a simple solution. It used videodiscs to allow the user to drive down corridors or streets, as if the vehicle were located in those corridors or on those streets. As our test case, we chose Aspen, Colorado (risking the Fleece), where the city's grid and size were manageable and where the townsfolk were sufficiently odd that they didn't worry about a homemade film truck driving down the middle of all the streets for several weeks, during several seasons.

The way the system worked was simple. Every street was filmed, in each direction, by taking a frame every three feet. Similarly, every turn was filmed in both directions. By putting the straight street segments on one videodisc and the curves on the other, the computer could give you a seamless driving experience. As you approached an intersection on, say, disc player 1, player 2 would line itself up at the intersection, and in the event that you decided to turn right or left, it would play that segment. While playing the turn, player 1 would then be free to seek out the straight segment of street onto which you had turned and, once again, would seamlessly play it as you ended your turn and started down the new street.

In 1978 the Aspen Project was magic. You could look out your side window, stop in front of a building (like the police station), go inside, have a conversation with the police chief, dial in different seasons, see buildings as they were forty years before, get guided tours, helicopter over maps, turn the city into animation, join a bar scene, and leave a trail like Ariadne's thread to help you get back to where you started. *Multimedia* was born.

The project was so successful that military contractors were hired to build working prototypes for the field, with the idea of protecting airports and embassies against terrorists. Ironically, one of the first sites to be commissioned was Tehran. Alas, it was not done soon enough.

BETA OF THE '90S

Today, multimedia offerings are mostly consumer products that, in the form of CD-ROM titles, have reached most Americans between the ages of five and ten, and an increasing number of adults as well. In 1994 more than two thousand CD-ROM consumer titles were available in the United States for the Christmas season. The current world population of all types of CD-ROM titles is estimated to be more than ten thousand. In 1995 almost every desktop computer shipped will have a CD-ROM drive in it.

A CD used as read-only memory (ROM) has a storage capacity today of 5 billion bits (using only one side, because that is easier to manufacture). This capacity will be increased to 50 billion on one side within the next couple of years. Meanwhile, 5 billion alone is huge, when you consider that an issue of the *Wall Street Journal* has approximately 10 million bits (thus, a CD-ROM can hold about two years' worth). Put another way, a CD represents about 100 classics or five years of reading, even for those who read two novels a week.

From another point of view, 5 billion is not so large; it is only one hour of compressed video. In this regard, the size is modest at best. One likely short-term result is that CD-ROM titles will use a lot of text—which is economically bitwise—many stills, some sound, and only snippets of full-motion video. Ironically, CD-ROMs may thus make us read more, not less.

The longer-term view of multimedia, however, is not based on that fifty-cent piece of plastic, 5 billion or 50 billion bits, but will be built out of the growing base of on-line systems that are effectively limitless in their capacity. Louis Rossetto, the founder of *Wired,* calls CD-ROMs the "Beta of the '90s," referring to the now-defunct Betamax video standard. He is certainly correct that, in the long term, multimedia will be predominantly an on-line phenomenon. Whereas the economic models for being on-line and for having your own CD-ROM may be different, with broadband access the functionality can be viewed as the same.

Either way, a fundamental editorial change takes place, because depth and breadth are no longer either/or. When you buy a printed encyclopedia, world atlas, or book on the animal kingdom, you expect very general and broad coverage of many far-ranging topics. By contrast, when you buy a book on William Tell, the Aleutian Islands, or kangaroos, you expect an "in depth" treatment of the person, place, or animal. In the world of atoms, physical limits preclude having both breadth and depth in the same volume—unless it's a book that's a mile thick.

In the digital world, the depth/breadth problem disappears and we can expect readers and authors to move more freely between generalities and specifics. In fact, the notion of "tell me more" is very much part of multimedia and at the root of hypermedia.

BOOKS WITHOUT PAGES

Hypermedia is an extension of hypertext, a term for highly interconnected narrative, or linked information. The idea came from early experiments at the Stanford Research Institute by Douglas Englebart and derived its name from work at Brown University by Ted Nelson, circa 1965. In a printed book, sentences, paragraphs, pages, and chapters follow one another in an order determined not only by the author but also by the physical and sequential construct of the book itself. While a book may be randomly accessible and your eyes may browse quite haphazardly, it is nonetheless forever fixed by the confines of three physical dimensions.

In the digital world, this is not the case. Information space is by no means limited to three dimensions. An expression of an idea or train of thought can include a multidimensional network of pointers to further elaborations or arguments, which can be invoked or ignored. The structure of the text should be imagined like a complex molecular model. Chunks of information can be reordered, sentences expanded, and words given definitions on the spot (something I hope you have not needed too often in this book). These linkages can be embedded either by the author at "publishing" time or later by readers over time.

Think of hypermedia as a collection of elastic messages that can stretch and shrink in accordance with the reader's actions. Ideas can be opened and analyzed at multiple levels of detail. The best paper equivalent I can think of is an Advent calendar. But when you open the little electronic (versus paper) doors, you may see a different story line depending on the situation or, like barbershop mirrors, an image within an image within an image.

Interaction is implicit in all multimedia. If the intended experience were passive, then closed-captioned television and subtitled movies would fit the definition of video, audio, and data combined.

Multimedia products include both interactive television and video-enabled computers. As discussed earlier, the difference between these two is thin, thinning, and eventually will be nonexistent. Many people (especially parents) think of "interactive video" in terms of Nintendo, Sega, and other makers of "twitch" games. Some electronic games can be so physically demanding that one has to get into a jogging suit in order to participate. The TV of the future, however, will not necessarily require the hyperactivity of Road Runner or the physique of Jane Fonda.

Today, multimedia is a desktop or living room experience, because the apparatus is so clunky. Even laptops, with their clamshell design, do not lend themselves to being very personal information appliances. This will change dramatically with small, bright, thin, flexible high-resolution displays. Multimedia will become more book-like, something with which you can curl up in bed and either have a conversation or be told a story. Multimedia will someday be as subtle and rich as the feel of paper and the smell of leather.

It is important to think of multimedia as more than a private world's fair or "son et lumière" of information, mixing fixed chunks of video, audio, and data. Translating freely from one to the other is really where the field of multimedia is headed.

MEDIUMLESSNESS

The medium is not the message in a digital world. It is an embodiment of it. A message might have several embodiments automatically derivable from the same data. In the future, the broadcaster will send out one stream of bits, like the weather example, which can be converted by the receiver in many different ways. The same bits can be looked at by the viewer from many perspectives. Take a sporting event, for example.

The incoming football bits can be converted by the computer-TV for you to experience them as a video; to hear them through an announcer; or to see them as diagrams of the plays. In each case it is the same game and same pool of bits. When those bits are turned into audio-only, the acoustic medium forces you to imagine the action (but allows you to drive a car at the same time). When the bits are turned into video, less is left to the imagination, but tactics are hard to see (because of the pell-mell or the sight of people piled on top of one another). When the bits are rendered as a diagram, the strategy and defense are quickly revealed. Moving among the three will be likely.

Think of a CD-ROM title on entomology as another example. Its structure will be more that of a theme park than a book. It is explored by different people in different ways. The architecture of a mosquito might best be represented in line drawings, its flight by animation, and its noise (obviously) through sound. But each incarnation need not be a different database or a separately crafted multimedia experience. They all could emanate from a single representation or be translated from one medium into another.

Thinking about multimedia needs to include ideas about the fluid movement from one medium to the next, saying the same thing in different ways, calling upon one human sense or another: if you did not understand it when I said it the first time, let me (the machine) show it to you as a cartoon or 3-D diagram. This kind of media movement can include anything from movies that explain themselves with text to books with gentle voice to read themselves to you out loud as you are dozing off.

A recent breakthrough in such automatic translation from one medium into another is the work of Walter Bender and his students at the Media Lab called "salient stills." The question they posed was, How could many seconds of video be printed as a still in such a way that the resolution of the still image would be an order of magnitude greater than any one frame? A single frame of 8mm video has very low resolution (two-hundred-plus lines) in comparison to a 35mm slide (thousands). The answer was to pull resolution out of time and look at many frames both forward and backward in time.

The research resulted in a process that can make very high-quality video prints (literally a three-foot-by-four-foot Kodacolor print) from crummy 8mm video. These stills have in excess of five thousand lines of resolution. This means that selections from the billions of hours of 8mm home movies stored in the shoe boxes of American homes can be turned into a portrait or Christmas card or printed for a photo album with as much or more resolution as a normal 35mm snapshot. Breaking news stories can be printed from CNN footage onto the front page of your newspaper or the cover of *Time* magazine, without recourse to those coarse images we sometimes see that make the world look like it is being seen through a blurry ventilator grill.

A salient still actually is an image that never existed. It represents a still frame of many seconds. During that time the camera may have zoomed and panned, and objects in the scene may have moved. The image is nonetheless crisp, without blur, and perfectly resolved. The still photo's contents reflect to some real degree the filmmaker's intentions by putting more resolution in places where the camera zoomed or by widening the scene if it panned. In Bender's method, quickly moving elements, like a person walking across a stage, drop out in favor of the temporarily stable ones.

This example of "multimedia" involves translating one dimension (time) into another dimension (space). A simple example is when a speech (the acoustic domain) is transcribed to print (the text domain) with punctuation indicating some intonation. Or, the script for a play in which the spoken lines are accompanied by many stage directions to establish the desired tone. These are forms of multimedia that often go unnoticed, but they, too, are part of a very large business.

THE BIT BUSINESS   6

A TWO-BIT STORY

I think of myself as an extremist when it comes to predicting and initiating change. Nonetheless, when it comes to technological and regulatory changes, as well as new services, things are moving faster than even I can believe—there is obviously no speed limit on the electronic highway. It's like driving on the autobahn at 160 kph. Just as I realize the speed I'm going, *zzzwoom,* a Mercedes passes, then another, and another. Yikes, they must be driving at 120 mph. Such is life in the fast lane of the infobahn.

Although the rate of change is faster than ever, innovation is paced less by scientific breakthroughs like the transistor, microprocessor, or optical fiber and more by new applications like mobile computing, global networks, and multimedia. This is partly because of the phenomenal costs associated with the fabrication facilities for modern chips, for which new applications are sorely needed to consume all that computing power and memory, and also because, in many areas of hardware, we are coming close to physical limits.

It takes about a billionth of a second for light to travel one foot, which is something not likely to change. As we make computer chips smaller and smaller, their speed can increase a little. But in order to make a big difference in overall computer power, it will be necessary to design new solutions, for example, with many machines running at the same time. The big changes in computers and telecommunications now emanate from the applications, from basic human needs rather than from basic material sciences. This observation has not gone unnoticed by Wall Street.

Bob Lucky, a highly acclaimed author and engineer and the vice president for applied research at Bellcore (formerly the exclusive research arm of the seven Baby Bells), noted recently that he no longer keeps up to date technically by reading scholarly publications; instead he reads the *Wall Street Journal*. One of the best ways to focus on the future of the "bit" industry is to set the tripod of one's telescope on the entrepreneurial, business, and regulatory landscape of the United States, with one leg each in the New York, American, and NASDAQ exchanges.

When QVC and Viacom battled for Paramount, analysts proclaimed the winner to be the loser. Paramount's financial performance did decline after the courtship started, but remains, nonetheless, a beautiful catch for Viacom because now it owns a wider variety of bits. Both Sumner Redstone and Barry Diller know that if your company makes only one kind of bit, you are not in very good shape for the future. The Paramount story was about bits, not egos.

The valuation of a bit is determined in large part by its ability to be used over and over again. In this regard, a Mickey Mouse bit is probably worth a lot more than a Forrest Gump bit; Mickey's bits even come in lollipops (consumable atoms). More interestingly, Disney's guaranteed audience is refueled at a rate that exceeds 12,500 births each hour. In 1994 the market value of Disney was $2 billion greater than that of Bell Atlantic, in spite of Bell Atlantic's sales being 50 percent greater and profits being double.

TRANSPORTING BITS

Transporting bits is an even worse business to be in than that of the airlines with their fare wars. The telecommunications business is regulated to such a degree that NYNEX must put telephone booths in the darkest corners of Brooklyn (where they last all of forty-eight hours), while its unregulated competitors will only put their telephone booths on Fifth and Park Avenues and in airline club lounges.

Worse, the entire economic model of pricing in telecommunications is about to fall apart. Today's tariffs are determined per minute, per mile, or per bit, all three of which are rapidly becoming bogus measures. The system is being ruptured by the wild extremes of time (a microsecond to a day), distance (a few feet to fifty thousand miles), and numbers of bits (one to 20 billion). In days when those differences were not so extreme, the old model worked well enough. When you used a 9600-bps modem, you paid 75 percent less for connect time than you did with your 2400-bps modem. Who cared?

But now the spread is huge, and we do care. Time is an example. Ignoring the transmission speed and the number of bits, am I to believe that I will be paying the same price to see a two-hour movie as I will be paying to have thirty different four-minute conversations? If I can deliver a fax at 1.2 million bps, am I really going to pay 1/125 the cost of what I pay today? If I can piggyback 16,000 bps voice on an ADSL movie channel, am I really going to pay five cents for a two-hour conversation? If my mother-in-law returns home from the hospital with a remotely monitored pacemaker that needs an open line to the hospital to monitor a half-dozen randomly spaced bits each hour, should those bits be tariffed the same way as the 12 billion bits in *Gone With the Wind*? Try figuring out that business model!

We have to evolve a more intelligent scheme. It may not use time, distance, or bits as the controlling variable and basis for tariff. Maybe bandwidth should be free, and we buy movies, long-distance health monitoring, and documents because of their value, not the channel's. It would be unconscionable to think of buying toys based on the number of atoms in them. It is time to understand what the bits and atoms mean.

If the management of a telecommunications company limits its long-term strategy to carrying bits, it will not be acting in its shareholders' best interest. Owning the bits or rights to the bits, or adding significant value to the bits, must be a part of the equation. Otherwise, there will be no place to add revenue, and telephone companies will be stuck with a service fast becoming a commodity, the price of which will go down further and further because of competition and increased bandwidth. But there is a catch.

When I was growing up, everyone hated the telephone company (as an adult I would put insurance companies at the top of the list). Any cunning 1950s child would have a scheme or a scam, and it was considered almost sporting to rip off the telephone company. Today, cable companies have assumed this honor, because many have provided poor service while raising their rates. Worse, the cable companies are not "common carriers"; they control what goes down their lines.

The cable industry has enjoyed many of the benefits of an unregulated monopoly, originally intended to be little more than a patchwork of community services. As cable franchises started to fuse and become national networks, people took notice of the fact that these companies indeed controlled both the telecommunications channel and the content. Unlike the telephone company, they were not obliged to provide a right-of-way except for very local and communal purposes.

Regulating the telephone industry is based on a simple principle: everybody is allowed to use it. But it is not clear what happens in a broadband system if it is more like today's cable companies than a telephone network. Congress is nervous about the fairhandedness with which a channel owner will welcome a content owner, given a choice. Also, if you own both content and channel, can you maintain your neutrality?

To put it another way, if AT&T and Disney merge, will the new company make it less expensive for children to access Mickey Mouse than Bugs Bunny?

GREENER BITS

In the fall of 1993, when Bell Atlantic agreed to buy cable giant Tele-Communications Inc. for $21.4 billion, pundits of the *information superhighway* took it as a signal that the digital age had truly begun. The digital ribbon had been cut.

However, the merger flew in the face of regulatory logic and common sense. Telephony and cable had positioned themselves as archrivals, regulation precluded most joint ownership, and loops and stars were thought to mix like oil and vinegar. Just the sheer level of investment dropped jaws.

Four months later, when the Bell Atlantic/TCI discussions collapsed, the pendulum overcompensated, and new jargon emerged about "roadkill" and "construction delays" on the information superhighway. The digital age had suddenly been postponed again, TCI's stock dropped more than 30 percent, and other associated companies took a spill as well. The champagne had to go back into the bottle.

But from my point of view, this was not an important casualty. In fact, the Bell Atlantic and TCI agreement was one of the least interesting corporate mergers. It was as if two plumbing-supply houses, marketing two distinctly different pipe sizes, had decided to combine their inventory. It was really not about the deep-rooted combination of channel and content, blending bit manufacturing with bit distribution. Disney and Hollywood king Michael Ovitz each teaming up in 1994 with three regional telephone companies—now that's more interesting.

Consumer electronics companies have tried to do this with entertainment companies. In principle, the idea is very powerful, but so far has had little synergy because of cultural differences of all kinds. When Sony bought CBS Records and then Columbia Pictures, Americans cried foul. Like the sale of Rockefeller Center, these purchases raised the issue of symbolic and real foreign control of a national cultural asset. When Matsushita bought MCA a while later, people were even more startled, because MCA's chairman, Lew Wasserman, was considered by many to be the most American chief executive. I remember visiting MCA headquarters after the first oil crisis and seeing stickers on the elevator buttons (a message from Lew) that said, "Walk up one and down two, for your health and your country." These purchases bring up deep cultural divides, not just between Japanese and American thought but between engineering and the arts. So far they have not worked, but I suspect they will.

CULTURE CONVERGENCE

There is a perceived polarity (however artificial) between technology and the humanities, between science and art, between right brain and left. The burgeoning field of multimedia is likely to be one of those disciplines, like architecture, that bridges the gap.

Television was invented through purely technological imperatives. When pioneers like Philo Farnsworth and Vladimir Zworykin looked at postage-stamp-size electronic images in 1929, they were driven to perfect the technology purely on the basis of its own merit. While Zworykin had some naive ideas about the use of television in the early days, he was sadly disappointed in his later years.

Former MIT president Jerome Wiesner tells a story about Zworykin's visiting him one Saturday at the White House when Wiesner was JFK's science advisor (and close friend). He asked Zworykin if he had ever met the president. As Zworykin had not, Wiesner took him across the hall to meet JFK. Wiesner introduced his visitor to the president as "the man who got you elected." Startled, JFK asked, "How is that?" Wiesner explained, "This is the man who invented television." JFK replied how that was a terrific and important thing to have done. Zworykin wryly commented, "Have you seen television recently?"

Technological imperatives—and only those imperatives—drove the development of television. Then it was handed off to a body of creative talent, with different values, from a different intellectual subculture.

Photography, on the other hand, was invented by photographers. The people who perfected photographic technology did so for expressive purposes, fine-tuning their techniques to meet the needs of their art, just as authors invented romance novels, essays, and comic books to fit their ideas.

Personal computers have moved computer science away from the purely technical imperative and are now evolving more like photography. Computing is no longer the exclusive realm of military, government, and big business. It is being channeled directly into the hands of very creative individuals at all levels of society, becoming the means for creative expression in both its use and development. The means and messages of multimedia will become a blend of technical and artistic achievement. Consumer products will be the driving force.

The electronic-games business ($15 billion worldwide) is an example. These games represent a business that is larger than the American motion-picture industry and growing much faster as well. Games companies are driving display technology so hard that virtual reality will become a "reality" at very low cost, whereas NASA was able to use it with only marginal success at a cost of more than $200,000. On November 15, 1994, Nintendo announced a $199 virtual reality game called "Virtual Boy."

Consider today's fastest Intel processor, which runs at 100 million instructions per second (MIPS). Compare that to Sony, which just introduced a $200 "Playstation" with a 1000 MIPS for the games market. What is going on? The answer is simple: our thirst for new kinds of entertainment is seemingly unquenchable, and the new, real-time 3-D content, which the games industry is banking on, needs that kind of processing and those new displays. The application is the imperative.

PULLING VERSUS PUSHING

Many of the big media companies like Viacom, News Corporation, and the publisher of this book add most of their value to information and entertainment content in one way: distribution. As I said earlier, the distribution of atoms is far more complex than of bits and requires the force of an enormous company. Moving bits, by contrast, is far simpler and, in principle, precludes the need for these giant corporations. Almost.

It was through *The New York Times* that I came to know and enjoy the writing of the computer and communications business reporter, John Markoff. Without *The New York Times*, I would never have known of his work. However, now that I do, it would be far easier for me to have an automatic method to collect any new story Markoff writes and drop it into my personalized newspaper or suggested-reading file. I would probably be willing to pay Markoff the proverbial "two cents" for each of his stories.

If one two-hundredth of the 1995 Internet population were to subscribe to this idea and John were to write a hundred stories a year (he actually writes between one-hundred-twenty and one-hundred-forty), he would earn $1,000,000 per year, which I am prepared to guess is more than *The New York Times* pays him. If you think one two-hundredth is too big a proportion, then wait a short while. The numbers really do work. Once somebody is established, the added value of a distributor is less and less in a digital world.

The distribution and movement of bits must also include filtering and selection processes. The media company is, among other things, a talent scout, and its distribution channel provides a test bed for public opinion. But after a certain point, the author may not need this forum. In the digital age, Michael Crichton could surely make far more money selling his next books direct. Sorry, Knopf.

Being digital will change the nature of mass media from a process of pushing bits at people to one of allowing people (or their computers) to pull at them. This is a radical change, because our entire concept of media is one of successive layers of filtering, which reduce information and entertainment to a collection of "top stories" or "best-sellers" to be thrown at different "audiences." As media companies go more and more toward narrowcasting, like the magazine business, they are still pushing bits at a special-interest group, like car fanatics, Alpine skiers, or wine enthusiasts. I recently encountered the idea of a niche magazine for insomniacs, which cleverly would advertise on late-night television, when the rates are low.

The information industry will become more of a boutique business. Its marketplace is the global information highway. The customers will be people and their computer agents. Is the digital marketplace real? Yes, but only if the interface between people and their computers improves to the point where talking to your computer is as easy as talking to another human being.

interface

WHERE PEOPLE AND BITS MEET

FATAL REACTION

I spend a minimum of three hours a day in front of a computer and have done so for many years, and I still find it very frustrating at times. Understanding computers is about as easy as understanding a bank statement. Why do computers (and bank statements) have to be so needlessly complicated? Why is "being digital" so hard?

They don't, and it need not be. The evolution of computing has been so fast that we've only recently had enough low-cost computing power to spend it freely on improving the ease of interaction between you and your computer. It used to be considered wasteful and frivolous to devote time and money to the user interface, because computer cycles were so precious and had to be expended on the problem, not the person.

Scientists would justify stoic interfaces in many ways. In the early 1970s, for example, a handful of "scholarly" papers were published on why black-and-white displays were "better" than color. Color is not bad. The research community just wanted to vindicate its inability to deliver a good interface at reasonable cost or, to be a bit more cynical, at the expense of some imagination.

Those of us working on the human-computer interface in the late 1960s and throughout the 1970s were considered computer sissies and eyed with outright contempt. Our work was not the right stuff, even though the field was gaining acceptance. To remind yourself of how important sensing, effecting, and feedback can be, think of the last time you pushed an elevator button and the light did not come on (presumably because the bulb was burned out). The frustration is enormous: did it hear me? Interface design and function are very important.

In 1972 there were only 150,000 computers in the world, whereas five years from now, the integrated-circuit manufacturer Intel alone expects to be shipping 100 million each year (and I think they are vastly underestimating). Using a computer thirty years ago, like piloting a moon lander, was the realm of a precious few schooled in the hocus-pocus needed to drive these machines, sometimes with primitive languages or none at all (just toggle switches and blinking lights). In my opinion, there was a subconscious effort to keep it mysterious, like the monopoly of the monks or some bizarre religious rite in the Dark Ages.

We are still paying the price today.

When people talk about the look and feel of computers, they are referring to the graphical user interface, which "professionals" call a GUI. The GUI improved enormously starting around 1971 with work at Xerox and, shortly after, at MIT and a few other places, and it culminated in a real product a decade later when Steve Jobs had the wisdom and perseverance to introduce the Macintosh. The Mac was a major step forward in the marketplace and, by comparison, almost nothing has happened since. It took all the other computer companies more than five years to copy Apple and, in some cases, they have done so with inferior results, even today.

The history of human endeavors to make machines more usable is almost exclusively devoted to enhancing the sensory points of contact and evolving better physical designs. The interface was treated largely as a traditional industrial design problem. The designers of teapots and rakes might consider the handle in terms of shape, heat transfer, and the prevention of blisters.

Cockpit design is a daunting challenge, not only because there are so many switches, knobs, dials, and gauges, but also because two or three sensory inputs of a similar kind can interfere with one another. In 1972 an Eastern Airlines L1011 crashed because its landing gear was not down. The voice of the air-traffic controller and the beeping of the on-board computer resulted in the crew not hearing the warning message. Deadly interface design.

At home I used to have a very intelligent VCR with near-perfect voice recognition and knowledge of me. I could ask it to record programs by name and, in some cases, even assume it would do so automatically, without my asking. Then, all of a sudden, my son went to college.

I have not recorded a TV program in more than six years. Not because I can't. It is because the value is too low for the effort. It is needlessly hard. More important, VCR usage and remote-control units in general have been treated as a button-pushing problem. Likewise, the general interface with personal computers has been treated as a physical design problem. But interface is not just about the look and feel of a computer. It is about the creation of personality, the design of intelligence, and building machines that can recognize human expression.

A dog can recognize you from your gait more than one hundred yards away, whereas a computer does not even know you are there. Almost any pet can tell when you are angry, but a computer does not have a clue. Even puppies know when they have done wrong; computers don't.

The challenge for the next decade is not just to give people bigger screens, better sound quality, and easier-to-use graphical input devices. It is to make computers that know you, learn about your needs, and understand verbal and nonverbal languages. A computer should know the difference between your saying "Kissinger" and "kissing her," not because it can find the small acoustic difference, but because it can understand the meaning. That's good interface design.

The burden of interaction today has been placed totally on the shoulders of the human party. Something as banal as printing a computer file can be a debilitating exercise that resembles voodoo more than respectable human behavior. As a result, many adults are turned off and claim to be hopelessly computer illiterate.

This will change.

ODYSSEYS

In 1968 Arthur C. Clarke shared an Oscar nomination with Stanley Kubrick for the movie *2001: A Space Odyssey.* Oddly, the movie came out before the book. Clarke was able to revise his manuscript after viewing the rushes (based upon an earlier version of the story). In a very real sense, Clarke was able to simulate his story line and refine his concepts. He was able to see and hear his ideas before committing them to print.

This may explain why HAL, the film's starring computer, was such a brilliant (if lethal) vision of a future human-computer interface. HAL (whose name does not come from the respective preceding letters of IBM) had a perfect command of speech (understanding and enunciating), excellent vision, and humor, which is the supreme test of intelligence.

Almost a quarter of a century passed before another example of interface excellence emerged: *The Knowledge Navigator.* This videotape, also a theatrical production, a so-called video prototype, was commissioned by the then CEO of Apple, John Sculley, whose own book was also called *Odyssey.* Sculley's book ended with ideas for a "knowledge navigator," later to become a video. He wanted to illustrate an interface of the future, beyond mice and menus. He did an excellent job.

The Knowledge Navigator depicts a flat book-like device open on the desk of a tweedy-looking professor. In one corner of its display is a bow-tied person who turns out to be the persona of the machine. The professor asks this agent to assist him in preparing for a lecture, delegates a handful of tasks, and on a couple of occasions is reminded of other matters. The agent can see, hear, and respond intelligently, like any human assistant.

What HAL and the Knowledge Navigator have in common is that they exhibit intelligence to such a degree that the physical interface itself almost goes away. Therein lies the secret to interface design: make it go away. When you meet somebody for the first time, you may be very conscious of their looks, speech, and gestures. But quickly the content of your communication dominates, even if it is largely expressed through tone of voice or the language of facial expressions. A good computer interface should behave similarly. The problem is less like designing a dashboard and more like designing a human.

Most interface designers, on the other hand, have been stubbornly trying to make dumb machines easier to use by smart people. They have taken their lead from the field called "human factors" in the United States and "ergonomics" in Europe, which is about how the human body uses its sensors and effectors to work with tools in its immediate surroundings.

A telephone handset is probably the most redesigned and overdesigned appliance on earth, yet remains utterly unsatisfactory. Cellular telephones make VCRs pale with their unusable interface. A Bang & Olufsen telephone is sculpture, not telephony, harder not easier to use than an old black rotary telephone.

Worse, telephone designs have been "featured" to death. Number storing, redialing, credit card management, call waiting, call forwarding, autoanswering, number screening, and on and on are constantly being squeezed onto the real estate of a thin appliance that fits in the palm of your hand, making it virtually impossible to use.

Not only do I not want all those features; I don't want to dial the telephone at all. Why can't telephone designers understand that none of us want to dial telephones? *We want to reach people on the telephone!*

Given half a chance, we would delegate that task, which suggests to me that the problem of a telephone may not be in the design of the handset, but in the design of a robot secretary that can fit in your pocket.

BEYOND ETCH-A-SKETCH

Computer interface design started in March 1960, when J. C. R. Licklider published his paper "Man-Computer Symbiosis." Lick (as he was called) was an experimental psychologist and acoustician by training who became a convert to and a messiah for computing, leading ARPA's initial computer efforts. He was asked in the mid-1960s to write an appendix for the Carnegie Commission report on the future of television. It was in that appendix that Lick coined the term *narrowcasting*. What Lick did not know at the time was that these two contributions, man-computer symbiosis and narrowcasting, were destined to converge in the 1990s.

Very early human-computer interface research, in the beginning of the 1960s, split into two parts that were not to rejoin for twenty years. One addressed interactivity, and the other focused on sensory richness.

Interactivity was tackled by solving the problem of sharing a computer, then an expensive and monolithic resource. In the 1950s and early 1960s, a computer was so valuable that you made every effort to keep it running nonstop. It was unthinkable to connect a keyboard and have, for example, the computer type out a question and then have it sit there idly while a human read, thought about, and replied to it. The invention, called time sharing, was a method by which multiple users could share a single machine from remote locations. If you divided the resource among, say, ten people, it was not just that each person could have one-tenth of the machine, but that one person's moment of reflection could be another person's full use of the computer.

Such sharing of the digital pie worked under the condition that no single user was a hog, needing vast amounts of computation or bandwidth. Early terminals ran at 110 baud. I remember vividly when that was increased to 300 baud; it seemed so fast.

By contrast, sensory richness was addressed with very high bandwidth graphical interaction. Early computer graphics required a machine fully dedicated to providing the image. It was no different in principle than today's personal computer, but it filled a large room and cost millions of dollars. Computer graphics was born as a line-drawing medium that demanded a great deal of computing power to control directly the beam of the cathode ray tube.

Only ten years later did computer graphics start to move away from a graphics of line drawing to a graphics of shapes and images. These new displays, called raster scan displays, required a lot of memory to store the picture point by point. They are so common today that most people don't know that they were originally considered heretical. (Almost nobody believed in 1970 that computer memory would ever be cheap enough to devote so much of it to graphics.)

Time sharing and computer graphics were poor bedfellows for the next two decades. Sensory-impoverished time-sharing systems emerged as the accepted tool for business and academic computing, giving birth to electronic banking and airline reservation systems, which we take for granted today. Commercial time sharing went hand in hand with very parsimonious interface design, usually typewriter output, almost purposely wanting the system to be slow enough for any single user, so others could get their fair share.

Computer graphics, on the other hand, developed for the most part with stand-alone computing. By 1968 so-called mini-computers in the $20,000 range started to emerge, largely because factory and machine automation needed very precise and real-time controls. So did computer graphics. Coupled with display devices, these stand-alone computer graphics systems were precursors to what we know today as workstations, which are nothing more than personal computers with long pants.

MULTIMODAL INTERFACE

Redundancy is generally considered a bad symptom, implying needless verbosity or careless repetition. In early human interface design, people studied techniques for interaction and tried to select judiciously one means or the other, for one set of circumstances or the other. Was a light pen better than a data tablet? The "either/or" mentality was driven by the false belief that there was a universal "best" solution for any given situation; it is false because people are different, situations change, and the circumstances of a particular interaction may well be driven by the channel you have available. There is no best in interface design.

I recall visiting an admiral in the mid-1970s who had one of the most advanced command-and-control systems. He would bark orders to a junior seaman, who would dutifully type in the proper commands. So, in this sense, the system had a terrific interface: it had speech-recognition facilities, and patience as well. The admiral could walk around the room, talk, and gesture. He could be himself.

Nonetheless, the admiral was unprepared to plan an attack through such an indirect interface. He knew the seaman was looking at the situation through the keyhole of the computer system's small display. The admiral preferred interacting directly with a large wall map of the "theater" on which he would thumbtack little blue and red ships of appropriate shape. (At the time we always joked about how the Russians were using the same colors.)

The admiral was comfortable with the map, not because it was old-fashioned and very high resolution but because he engaged with it with his whole body. As he moved ships, his gestures and motor actions reinforced his memory. He was deeply involved with the display, right down to his neck muscles. It was not an either/or interface; it was both/and.

Both/and led to a breakthrough in thinking, which, simply stated, said that redundancy is good. In fact, the best interface would have many different and concurrent channels of communication, through which a user could express and cull meaning from a number of different sensory devices (the user's and the machine's). Or, equally important, one channel of communication might provide the information missing in the other.

For example, if we are in a room with a handful of people, and I say to somebody, "What is your name?" the question really does not have any meaning unless you can see where I am looking while I speak. Namely, the adjective *your* gets its meaning from the direction of my gaze.

This was beautifully illustrated in a program called Put-That-There, developed at MIT by Dick Bolt and Chris Schmandt. The first embodiment of the program, in 1980, allowed you to speak and gesture at a wall-size display and move simple objects (later ships) around a blank screen (later the Caribbean). In a filmed demonstration of Put-That-There, it misrecognized a command. Schmandt's spontaneous "Oh, shit" was engraved in film stock to remind many future audiences of how much work was still to be done.

The idea is simple: talking, pointing, and looking should work together as part of a multimodal interface that is less about messaging back and forth (the basis of time sharing) and more like face-to-face, human-to-human conversation.

At the time, this and other early attempts at an alternate "both/and" approach to the interface design looked like sloppy science. I have little respect for testing and evaluation in interface research. My argument, perhaps arrogant, is that if you have to test something carefully to see the difference it makes, then it is not making enough of a difference in the first place.

THE NOTICEABLE DIFFERENCE

When I was a little boy, my mother had a linen closet, the back of which had a "secret wall." The secret was no big deal: a collection of pencil lines that we periodically and carefully made to mark my height. All the pencil lines were dutifully dated, and some were close together due to frequency and others were spread apart because, for example, we had been away for the summer. Using two closets did not seem to make sense.

This scale was a private matter, and I guess in some way measured my intake of milk, spinach, and other good things.

By contrast, growth has a more dramatic face. A rarely seen uncle might comment, "How much you have grown, Nicky?" (since he had seen me two years prior, we suppose). But I could not really comprehend the change. All I could see were the little lines in the linen closet.

The "just-noticeable difference," or JND, is a unit of measure in psychophysics. Its name alone has influenced human interface design. You have to ask yourself, if it is a JND, why bother? If you have to measure carefully to see any difference at all, maybe we are not working on things that matter enough.

For example, scholarly studies have suggested that speech and natural language are not appropriate channels of communication between people and computers in most applications. These technical reports are filled with tables, control groups, and the like, proving that natural language is confusing for human-computer communication.

While I certainly don't expect the pilot of a 747 to taxi and take off by singing "Up, Up and Away," I still can't fathom any reason not to use the richness of speech and gesture, even in the cockpit. Wherever the computer may be, the most effective interface design results from combining the forces of sensory richness and machine intelligence.

When this happens, we'll see a noticeable difference. We'll see what my uncle saw, instead of the little lines in the closet.

INTELLIGENT INTERFACES

My dream for the interface is that computers will be more like people. This idea is vulnerable to criticism for being too romantic, vague, or unrealizable. If anything, I would criticize it for shooting too low. There may be many exotic channels of communications of which we may not even be aware today. (As somebody married to an identical twin and with identical twin younger brothers, I am fully prepared to believe from observation that extrasensory communication is not out of the question.)

In the mid-1960s, I set my goals by trying to emulate face-to-face communication, with its languages of gesture and facial expressions and the motor involvement of our body and limbs. I used the admiral as my model.

In a landmark project called the Spatial Data Management System (circa 1976), the goal was to provide a human interface that would "bring computers directly to generals, presidents of companies, and six-year-old children." The system was designed to be learnable in thirty seconds. Familiarity with desktops and bookshelves was the tool used to browse and manipulate complex audio, video, and data.

That was radical for the late 1970s, but it still missed the more meaningful consequence of fashioning our communication after the conversation between the admiral and the seaman. Future human-computer interface will be rooted in delegation, not the vernacular of direct manipulation—pull down, pop up, click—and mouse interfaces. "Ease of use" has been such a compelling goal that we sometimes forget that many people don't want to use the machine at all. They want to get something done.

What we today call "agent-based interfaces" will emerge as the dominant means by which computers and people talk with one another. There will be specific points in space and time where bits get converted into atoms and the reverse. Whether that is the transmission of a liquid crystal or the reverberation of a speech generator, the interface will need size, shape, color, tone of voice, and all the other sensory paraphernalia.

GRAPHICAL PERSONA ‎8

GRAPHICS' BIG BANG

At MIT in 1963, the Ph.D. thesis of Ivan Sutherland, called "Sketchpad," exploded upon the world the idea of interactive computer graphics. Sketchpad was a real-time line-drawing system that allowed the user to interact directly with the computer screen by means of a "light pen." The achievement was of such magnitude and breadth that it took some of us a decade to understand and appreciate all of its contributions. Sketchpad introduced many new concepts: dynamic graphics, visual simulation, constraint resolution, pen tracking, and a virtually infinite coordinate system, just to name a few. Sketchpad was the big bang of computer graphics.

During the next ten years many researchers seemed to lose interest in the real-time and interactive aspects of computer graphics. Instead, most creative energy went into synthesizing realistic images, off-line and not in real time. Sutherland himself was slightly detoured by the problem of visual verisimilitude, i.e., just how photorealistic and detailed he could make a computer image. Such problems as shadows, shading, reflections, refractions, and hidden surfaces were the focus of that research. Beautifully rendered chess pieces and teapots became the icons of the post-Sketchpad period.

During this same time, I came to believe that the comfort and ease with which humans could express their graphical ideas were more important than the machine's ability to render them as synthetic photographs. Good human-computer interface design included the computer understanding incomplete, ambiguous thoughts, typical of the early stages in any design process, versus the more complete and consistent presentations of complex, finished renderings. The on-line and real-time tracking of a hand-drawn sketch provided me with an excellent area for research into understanding and advancing computer graphics as a more dynamic, interactive, and expressive medium.

The key concept of my work was to understand a person's graphical "intent." If a user slowly drew a gentle and seemingly purposeful curve, the computer presumed he or she intended it as such, whereas the very same shaped line, if quickly drawn, may well have been intended to be a straight one. If these two gentle curves were viewed after the fact, instead of while they were drawn, they could look exactly alike. The user's drawing behavior, however, indicated two totally distinct differences in intent. Sketching behavior, moreover, varied from person to person, as we all draw differently. Therefore, the computer had to learn about the sketching style of each user. The same concept is found thirty years later in the Apple Newton's ability (which some question) to recognize handwriting by adapting to the user's penmanship (those who have given it time seem more satisfied).

Recognizing sketched shapes and objects led my thinking about computer graphics away from lines and more toward dots. In a sketch, what is between the lines or enclosed by them is the most important part for understanding what the drawing is about.

At Xerox's PARC, during the same period, researchers had also invented a shape-oriented approach to computer graphics in which amorphous areas were handled and textured by storing and displaying images as a massive collection of dots. A few of us in that period concluded that the future of interactive computer graphics was not in line-drawing engines like Sketchpad but in television-like raster scan systems, which mapped images (stored in computer memory) onto a display device, versus driving the beam of a CRT in X and Y, like an Etch-A-Sketch. The primitive element of computer graphics, which had been the line, now became the pixel.

THE POWER OF THE PIXEL

In the same way that a bit is the atomic element of information, a pixel is the molecular level of graphics. (I do not call it the atomic level, because a pixel is usually represented by more than one bit.) The computer graphics community invented the term *pixel,* which comes from the words *picture* and *element.*

Think of an image as a collection of rows and columns of pixels, like a crossword puzzle without entries. For any given monochrome image you can decide how many rows and columns you wish to use. The more of each, the smaller the squares, the finer the grain, and the better the result. In your mind, lay this grid over a photograph and fill each square with a value of the light intensity. The completed crossword puzzle will be an array of numbers.

In the case of color you have three numbers per pixel, usually one each for red, green, and blue or one each for intensity, hue, and saturation. Red, yellow, and blue, as we are taught in grade school, are not the three primary colors. The three additive primaries (i.e., in television) are red, green, and blue. The three subtractive primaries (as in printing) are magenta, cyan, and yellow. Not red, yellow, and blue. (I am told that we do not teach these terms to kids because the word *magenta* is too long. Many adults have never heard of *cyan.* Anyway.)

In the case of motion, time is sampled—as with the frames of a movie. Each sample is a single frame, another crossword puzzle, and when stacked together and played back in sufficiently rapid succession, the visual effect of smooth motion is produced. One of the reasons you have seen so few dynamic graphics or you have found video displayed in a small window is that it is hard to get sufficient numbers of bits out of memory and onto the screen as pixels fast enough (to produce the 60 to 90 frames a second needed for flicker-free smooth motion). As each day passes, somebody offers a new product or technique to do this faster.

The real power of the pixel comes from its molecular nature, in that a pixel can be part of anything, from text to lines to photographs. *Pixels are pixels* is as true as *bits are bits*. With enough pixels and with enough bits per pixel (for graytone or color), you can achieve the excellent display quality of contemporary personal computers and workstations. However, the bad qualities, just as much as the good ones, result from the strictures of this basic grid structure.

Pixels tend to require a lot of memory. The more pixels and the more bits per pixel you use, the more memory you need to store them. A typical screen with 1,000 by 1,000 pixels, in full color, needs 24 million bits of memory. When I was a freshman at MIT in 1961, memory cost about a dollar per bit. Today, 24 million bits costs $60, which means we can more or less ignore the large appetite for memory of pixel-oriented computer graphics.

As recently as five years ago, this was not the case, and people economized by using fewer pixels per screen and by using far fewer bits per pixel. In fact, early raster scan displays tended to use only one bit per pixel, out of which we inherited a special problem: the jaggies.

UNACCEPTABLE JAGGIES

Have you ever wondered why your computer screen has jagged lines? Why do images of pyramids look like ziggurats? Why do uppercase *E*, *L*, and *T* look so good, yet *S*, *W*, and *O* look like badly made Christmas ornaments? Why do curved lines look like they've been drawn by someone with palsy?

The reason is that only one bit per pixel is being used to display the image, and the result is a staircase effect, or spatial aliasing, which is absolutely unnecessary if hardware and software manufacturers would just use more bits per pixel and throw a little computing power at the problem.

So, why aren't all computer displays anti-aliased? The excuse given is that it takes too much computing. Ten years ago, one could accept the argument (maybe) that computer power was best spent elsewhere; in addition, the intermediate levels of gray needed to anti-alias were not as common as they are today.

Unfortunately, the consumer has been trained to accept the jaggies as a given. We even seem to be turning this artifact into a kind of mascot in much the same way that graphic designers frequently used that funny, magnetic-readable font, MICR, in the 1960s and 1970s to create an "electronic" look. In the 1980s and 1990s designers are doing this again by using exaggerated, aliased typography to mean "computerness." Today there is no need for lines and characters to be anything less than print quality and perfectly smooth. Don't let anybody tell you otherwise.

ICONOGRAPHICS

In 1976 Craig Fields, a program director in the Cybernetics Technology Office at ARPA (and later the director of ARPA itself), commissioned a New York computer animation company to produce a movie of a fictitious desert town called Dar El Marar. The animated movie depicted a cockpit view from a helicopter flying around Dar El Marar, swooping into its streets, pulling back to see the whole townscape, visiting neighborhoods, and moving in close to see into buildings. The movie simulated being Peter Pan, not for the purpose of experiencing the townscape and a world of buildings but for exploring a world of information. The concept assumed that you had designed the town; you had built neighborhoods of information by storing data in particular buildings, like a squirrel storing nuts. Later, you would retrieve the information on your magic carpet by going to where you had stored it.

Simonides of Ceos (556–468 B.C.) was a poet of classical Greece who was noted for his prodigious memory. When the roof of a banquet hall collapsed just after he had been called from the room, he found that he could identify the mangled remains of guests based on where they had been sitting: he inferred that tying material to specific spots in a mental spatial image would aid recall. He used this technique to remember his long speeches. He would associate parts of his oration with objects and places in a temple. Then while delivering his speech, he would revisit the temple in his mind to call forth his ideas in an orderly and comprehensive manner. The early Jesuits in China called this same process the building of "palaces of mind."

These examples involve navigating in three-dimensional space to store and retrieve information. Some people are good at this; some are not.

In two dimensions most of us are uniformly more capable. Consider the two-dimensional façade of your own bookshelves. You probably know how to find any book simply by going to "where" it is. You probably remember its size, color, thickness, and type of binding. You certainly recall this information much better if you put it "there." The messiest desktop is known to its user because that user made the mess, so to speak. There would be nothing worse than to have a librarian come in and reorganize your books by the Dewey decimal system, or a maid to arrive to clean up your desk. You would suddenly be lost.

Observations such as these led to the development of what we called a *spatial data management system*. SDMS was embodied in a room with a floor-to-ceiling, wall-to-wall, full-color display; two auxiliary desktop displays; octophonic sound; an instrumented Eames chair; and other paraphernalia. SDMS offered the user a sofa-style interface and the armchair opportunity to fly over data and gaze out of a picture-window-size display. The user could zoom and pan freely in order to navigate through a fictitious, two-dimensional landscape called Dataland. The user could visit personnel files, correspondence, electronic books, satellite maps, and a whole variety of totally new data types (like a video clip of Peter Falk in "Columbo" or a collection of fifty-four thousand still frames of art and architecture).

Dataland itself was a landscape of small images that illustrated the function or data behind them. Behind an image of a desktop calendar was the user's agenda. If the user drove the system into the image, for example, of a telephone, SDMS would initiate a telephoning program with associated personal Rolodex. This was the birth of icons. We toyed with using the word *glyphs* instead, because the dictionary meaning of *icons* is not really appropriate, but *icons* stuck.

These postage-stamp-size images not only illustrated data or functionality, but each had a "place." As with books on a bookshelf, you would retrieve something by going to where it was, remembering its location, color, size, and even the sounds it might make.

SDMS was so far ahead of its time that a decade had to pass and personal computers had to be born before some of the concepts could move into practice. Today, icons are common to the persona of all computers. People consider the imagery of trash cans, calculators, and telephone handsets as standard fare. In fact, some systems literally refer to the screen as a "desktop." What has changed is that today's Datalands are not spread floor to ceiling, wall to wall. Instead, they are accordioned into "windows."

THE SHAPE OF WINDOWS

I'm always impressed by how clever naming can scoop the market, leaving the consumer with very false impressions. When IBM chose to call its personal computer the PC, it was a stroke of genius. In spite of Apple's having been on the market more than four years earlier, the name *PC* has become synonymous with personal computing. Likewise, when Microsoft chose to name its second-generation operating system *Windows*, it brilliantly claimed the term forever after, in spite of Apple's having had (better) windows more than five years earlier and many workstation manufacturers already widely using them.

Windows exist because computer screens are small. The result is that a relatively small work space can be used to keep a number of different processes active at any one time. This whole book was written on a nine-inch diagonal screen with no paper, except that produced for or by the publisher. To most people, using windows is like riding a bicycle; you don't even remember learning how to do it, you just do it.

Windows are also interesting as a metaphor for the future of television. In the United States, more than in other countries, we have insisted in the past that a television image fill the screen fully. But filling the screen has a cost that derives from the fact that not all movies and television programs were created in the same rectangular format.

In fact, in the early 1950s the movie industry quite purposely moved to a number of wider-screen processes (such as Cinerama, Super Panavision, Super Technirama, 35mm Panavision, and Cinemascope, which we still use today) in order to undermine early television distribution. The 3-by-4 aspect ratio of today's television was derived from the pre–World War II generation of movies and does not fit Cinemascope or, for that matter, the rectangular format of most films produced in the past forty years.

Broadcasters in continental Europe resolve the difference in aspect ratio by so-called letter boxing. They blank the top and the bottom of the screen with black, so that the remaining, active area has the correct aspect ratio. By sacrificing a few pixels, the viewer gets to see the film in faithful replication of the shape of each frame. In fact, I think that the effect of letter boxing is additionally satisfying because it introduces a very crisp horizontal edge at the top and at the bottom of the image, which would otherwise be less sharply delimited by the curved plastic edge of the television set.

We rarely do this in the United States. Instead, we "pan-and-scan" when we transfer film to video, taking a wide-screen movie and collapsing it into a 3-by-4 rectangle. We don't just squash the picture (though that is done with titles and credits). Instead, during the transfer process, as the film is moving through the machine (usually a flying spot scanner), a human operator manually moves a 3-by-4 window over the much wider film window, sliding it one way or another, to catch the most relevant parts of each scene.

Some filmmakers, notably Woody Allen, will not allow this, but most seem not to care. One of the best examples of where such pan-and-scan failed hopelessly was in *The Graduate*. In the scene where Dustin Hoffman and Anne Bancroft are taking off their respective clothes, each at one extreme of the screen, there was no way for the operator to get both of them at the same time in the same frame of video.

In Japan and Europe there has been a major push for a new and wider aspect ratio of 9 by 16, and HDTV contestants in the United States are sheepishly following. Nine by sixteen may in fact be worse than 3 by 4, however, because all existing video material (which is 3 by 4) will now have to be displayed with vertical strips of black on each side of your 9-by-16 screen, so-called curtains. Not only do curtains serve less of a visual purpose than letter boxing; there is no way to pan-and-scan instead, even if you wanted to.

Aspect ratio should be a variable. When TVs have enough pixels, a windows style makes enormous sense. The ten-foot experience and the eighteen-inch experience start to collapse into one. In fact, in the future, when you have massively high resolution and a wall-size display, floor to ceiling and wall to wall, you may place your TV image on the screen as a function of where the plants are in the room, as opposed to the frame around some small screen. It's the whole wall.

CONSUMER GRAPHICS

As recently as five years ago, computer manufacturers, including Apple, were unwilling to pursue aggressively the home as a market. Hard to believe. A few years earlier, the stock price of Texas Instruments surged when it announced that it was dropping out of the home computer business.

In 1977 Frank Cary, chairman of IBM, announced to stockholders that IBM would enter the consumer electronics business. A task force was appointed in typical IBM style, and it reviewed a number of candidates, including, among others, wristwatches. IBM settled on the home computer. A top-secret project followed, code-named *Castle*, in which I participated as an advisor one day a week. A very ambitious personal computer was conceived, and it had a built-in digital videodisc.

The distinguished industrial designer Elliott Noyes created a prototype home computer that all of us would be proud to have in our houses twenty years later. But the dream started to unravel. IBM's labs in Poughkeepsie, New York, could not get the transmissive (the laser went through a transparent disc, as opposed to reflecting off a shiny one), floppy, ten-hour digital videodisc to work properly, so the personal computer and the videodisc were separated. *Castle* was divided.

The personal computer piece of the program was sent to another IBM lab, in Burlington, Vermont, and later to Boca Raton (the rest of that part is history). The videodisc was eventually scuttled in favor of a joint venture with MCA (something both companies regretted shortly afterward). *Castle* was stillborn, and the personal computer had to wait a few more years for Steve Jobs's garage.

At about the same time, electronic games introduced a different nature of computers and graphics. These consumer products were highly dynamic because of their intrinsic interactivity. In addition, their hardware and content blend so naturally. Games manufacturers do not make any money on the hardware; they make it on the games. It is truly a story about razors and blades.

But the games manufacturers, like those proprietary-minded computer companies that are now extinct, have so far missed the opportunity to open up their closed systems and compete with imagination. Sega and Nintendo will also be extinct if they do not wake up to the fact that PCs are eating their lunch.

The free-lance designers of games today must realize that their games are most likely to be best-sellers if built for a general-purpose platform, of which Intel alone plans to sells a hundred million a year. For this reason, the computer graphics of PCs will evolve rapidly toward what you see today in the most advanced arcade games. PC-based games will overtake dedicated game systems as we know them today. The only place where special-purpose hardware may play a near-term role is in virtual reality.

20/20 VR

OXYMORON OR PLEONASM

Mike Hammer (not the detective, but the world's leading corporate doctor or so-called re-engineer) calls *corporate change* an oxymoron on its way to becoming a pleonasm. A pleonasm is a redundant expression like *in one's own mind.* It arguably is the inverse of an oxymoron, which is an apparent contradiction like *artificial intelligence* or *airplane food.* If prizes were awarded for the best oxymorons, *virtual reality* would certainly be a winner.

If the component words of *virtual reality* are seen as *equal halves,* thinking about VR as a redundant concept makes more sense. VR can make the artificial as realistic as, and even more realistic than, the real.

For example, flight simulation, the most sophisticated and longest-standing application of VR, is more realistic than flying a real plane. Newly trained but fully able pilots take the controls of a completely loaded passenger plane for their first flight in a "real" 747, because they have learned more in the simulator than they could have learned in the actual plane. In the simulator, a pilot can be subjected to all sorts of rare situations that, in the real world, could be impossible, could require more than a near miss, or could rip apart an actual plane.

Another socially responsible application of VR would be to require its use in driving schools. On a slippery road, a child darts out from between two cars—none of us knows how we might react. VR allows one to experience a situation *with one's own body*.

The idea behind VR is to deliver a sense of "being there" by giving at least the eye what it would have received if it were there and, more important, to have the image change instantly as you change your point of view. Our perception of spatial reality is driven by various visual cues, like relative size, brightness, and angular movement. One of the strongest is perspective, which is particularly powerful in its binocular form in that your right and left eyes see different images. Fusing those images into one 3-D perception is the basis of stereovision.

The perception of depth provided by each eye seeing a slightly different image, eye parallax, is most effective for objects very near you (within, say, six feet). Objects farther away essentially cast the same image on each eye. Have you ever wondered why a 3-D movie has so much back-and-forth motion in the near field, with objects always flying into the audience? It is because that is where stereoscopic images work best.

The typical dress code for VR is a helmet with goggle-like displays, one for each eye. Each display delivers a slightly different perspective image of what you would see if you were there. As you move your head, the images are, in principle, so rapidly updated that you feel you are making these changes by moving your head (versus the computer actually following your movement, which it is). You feel you are the cause, not the effect.

The measure of just how real the visual experience may seem is a combination of two factors. One is image quality: the number of displayed edges and the textures between them. The other is response time: the speed with which these scenes are updated. Both variables are extremely demanding of computer power and, until recently, have been out of reach to most product developers.

VR started as far back as 1968, when none other than Ivan Sutherland built the first head-mounted display system. Later work at NASA and by the Department of Defense led to some expensive prototypes for space exploration and military applications. Tank and submarine trainers were particularly well-suited uses for VR, because the "real" experience had you looking into binoculars or a periscope anyway.

It is only now that we have sufficiently high-powered and low-costing computers to think of this technology as a consumer entertainment medium. In this new context, it will be nothing short of awesome.

THE COUCH COMMANDO

Jurassic Park would make a fabulous VR experience. Unlike the book or movie of the same name, it would not have a story line. Michael Crichton's job in this case would be that of stage-set or theme-park designer and the one who imbues each dinosaur with appearance, personality, behavior, and purpose. Put the simulation in motion. Then you enter. This is not television, and it need not be as antiseptic as Disneyland. There are no crowds, no queues, no popcorn smells (just dinosaur dung). It is like being in a prehistoric jungle and can be made to seem more dangerous than any real jungle.

Future generations of adults as well as children will entertain themselves in this manner. Since the imagery is computed, not real, there is no need to limit oneself to life-size or real places. VR will allow you to put your arms around the Milky Way, swim in the human bloodstream, or visit Alice in Wonderland.

VR today has drawbacks and technical failings that have to be corrected before the experience will have widespread appeal. For example, low-cost VR is plagued by aliased graphics. In the case of motion, the jaggies are even more unsettling because they appear to move and they don't necessarily move in the same direction as the scene. Think of the horizon line, perfectly horizontal. Now tilt it, just ever so slightly, so that one jagged step appears in the middle. Then tilt it a tiny bit more. Two appear, then three, then more; then they look as if they are moving, until you come to a forty-five-degree angle, where the line now is a perfect staircase of jaggies composed of pixels that abut corner to corner. Ugly.

Worse, VR is not yet fast enough. All commercial systems, particularly those that will soon be brought to you by the major videogame manufacturers, have a lag. As you move your head, the image changes rapidly, but not rapidly enough. The image lags.

In the early days of 3-D computer graphics, a variety of stereoscopic eyeglasses were used to achieve the effect. Sometimes these were inexpensive polarized lenses, and sometimes they were more expensive electronic shutters, which alternately exposed each eye to a different image.

I remember when I first worked with such devices, everybody—not most people, but literally everybody—after putting these glasses on for the first time and seeing 3-D on the screen, would then move their heads from side to side, looking for the images to change. As with 3-D movies, this did not happen. Head movement had no effect.

This human response, a "neck jerk" reaction, says it all. VR really needs to be tightly coupled with motion and position sensing to enable the viewer to cause change, not just the machine. In VR one must track the head, and responding to the rapidity of its change is almost all that counts. The speed with which the image is updated (the frequency response) is actually more important than resolution—an example of where our motor-sensory system is so acute that even the slightest delay ruins the experience.

Most manufacturers will probably miss the point totally and will market early VR systems that have as much image resolution as possible, at the expense of response time. They would be providing a much more satisfactory VR experience by displaying less graphics, anti-aliasing the images, and delivering rapid response.

The alternative is to abandon altogether head-mounted displays, which deliver perspective images to each eye separately, and move to so-called auto-stereoscopic technologies, which float a real object or holographic image in space, delivering it to both eyes.

TALKING HEADS

In the mid-1970s, ARPA launched a major research initiative in teleconferencing in order to address an important aspect of national security. The specific need was to transmit electronically the fullest possible sense of human presence for five particular people at five different sites. Each of these five people, physically separated, had to believe that the other four were physically present.

This extraordinary telecommunications requirement was driven by the government's emergency procedures in the event or threat of a nuclear attack. During the 1970s, the following action was taken: the president of the United States, the vice president, the secretary of state, the chairman of the Joint Chiefs of Staff, and the Speaker of the House immediately would go to a well-known site under a mountain in Virginia. There they would defend our nation from an advanced command-and-control room (like the one in the movie *WarGames*), supposedly immune to attack and fallout.

But just how safe was it to have all five people in a single known place? Wouldn't it be much safer to have them at five different locations (one in the air, one in a submarine, one under a mountain, etc.) if they could feel that they were together in one place? Clearly, yes. For this reason, ARPA funded advanced research in teleconferencing, through which my colleagues and I were awarded a contract to digitally create a real-time, human "telepresence."

Our solution was to replicate each person's head four times, with a life-size translucent mask in the exact shape of that person's face. Each mask was mounted on gimbals with two degrees of freedom, so it could nod and the head could turn. Perfectly registered video was projected inside of it.

Each site was composed of one real person and four bobbing plastic heads sitting around a table in the same order. Each person's video image and head position was captured and transmitted. If the president turned and spoke to the vice president, the secretary of state would see their respective plastic heads do so at his site. Admittedly bizarre.

Video thus projected resulted in lifelike emulation so vivid that one admiral told me that the "talking heads" gave him nightmares. He preferred an uppercase-only telegram on yellow paper from the president saying "FIRE," versus the commander-in-chief's bobbing head on the bridge of his aircraft carrier. His response is odd given his paranoia about whether the video image and voice were really those of the president (as opposed to somebody pretending to be the president). A telegram is much easier to fake.

We probably will not know how to decompose, transmit, and recompose people (or cheeseburgers or cashmere sweaters) for the next one or two millennia. But in the meantime, there will be many display techniques that will depart from the flat and near-flat screens to which we have become so accustomed. The surrounding edge of displays, the so-called bezel, will become less constricting for images large and small. Some of the most imaginative digital apparatuses of the future will have no edge at all.

R2D2 3-D

Sometime in the next millennium our grandchildren or great-grandchildren will watch a football game (if they call it that) by moving aside the coffee table (if they call it that) and letting eight-inch-high players run around the living room (if they call it that) passing a half-inch football back and forth. This model is the exact opposite of early VR thinking. All the resolution is provided everywhere for any point of view. Wherever you look, you see 3-D pixels (sometimes called voxels or boxels) floating in space.

In *Star Wars,* R2D2 beamed Princess Leia on Obi-Wan Kenobi's floor in this manner. The beautiful princess was a ghostly apparition projected in space, viewable (in principle) from any angle. This special effect, like so many similar ones in *Star Trek* and other science-fiction movies, has inadvertently created a blasé audience for technologies like holography. We have seen it so often in the movies; it is presumed to be easier than it is.

In fact, it took MIT's Professor Stephen Benton, who invented the white light hologram (common today on credit cards), more than twenty years to arrive at a similar result using the power of a million-dollar supercomputer, almost priceless special-purpose optics, and the relentless energy of a dozen brilliant Ph.D. students.

Holography was invented by the Hungarian scientist Dennis Gabor in 1948. In the most simple terms, a hologram is a collection of all possible views of a scene into a single plane of light modulating patterns. When light is later passed through or reflected from this plane, the scene is optically reconstructed in space.

Holography has been a dark horse in the race to make better and better displays. One of the reasons this is true is because holography requires massive resolution. Your TV is supposed to have 480 visible scan lines (it probably has far fewer). If you have a TV screen that measures a modest ten inches in height, that means (under the very best conditions) you have almost 50 lines per inch. Holography requires about 50,000 lines per inch, or a thousand times more horizontal scan lines. Worse, resolution is in both X and Y, so it is a thousand squared, or a million times the resolution of TV today. One of the reasons you see holograms on credit cards and even some countries' bank notes is that this resolution requires very sophisticated, hard-to-forge printing technology.

The reason that Benton and his colleagues have made any progress at all is that they cleverly figured out what the human eye and the human perceptual system really need, versus what a native hologram can produce. Since the human eye is the client for the image, it would be silly to present it with more detail than it could resolve. Similarly, he noted that we can look at spatial imaging (sampling in space) in much the same way we look at the individual frames in a movie (sampled in time). Video offers smooth motion at about 30 frames (60 fields) per second. Therefore, instead of making a hologram represent *every* point of view, why not represent a point of view every fraction of an inch and leave out the data in between? It works.

In addition, Benton and his colleagues noted that our sense of spatiality is largely horizontal. Because of side-by-side eye parallax and because we tend to move across mostly horizontal planes, horizontal parallax is the far more dominant spatial cue than vertical parallax (the up-down change). If our eyes were stacked one above the other, or if we climbed up and down trees a lot, this would not be the case. We don't. In fact, the horizontal emphasis of our perception is so strong that Benton concluded that he could discard vertical parallax altogether.

For this reason, almost none of the holograms displayed at the Media Lab have any vertical parallax. When I show people the small gallery of examples hanging outside Benton's lab, visitors just do not notice. In fact, once told, people will bend their knees and stand on their tiptoes a few times to really believe it.

The result of spatial sampling combined with using only horizontal parallax means that Benton's group needs about a ten-thousandth of the computing power required to make an otherwise fully resolved hologram. For this reason, they have achieved the world's first real-time holographic video of full-color, shaded images, which floats freely in space. It is about the size and shape of a teacup or a dumpy-looking Princess Leia.

MORE THAN MEETS THE EYE

The quality of a display is literally more than meets the eye. It is a viewing experience that typically engages other senses. The collective sensation as a whole is truly greater than the sum of the parts.

In the early days of HDTV, social scientist Russ Neuman, then at the Media Lab, conducted a landmark experiment on audience response to display quality. He set up two absolutely identical high-quality TVs and top-of-the-line VCRs, playing the exact same high-quality videocassettes. However, in the one setup (A), he used the normal sound quality of the VCR and the TV set's tiny speakers. In the other (B), he used better-than-CD-quality sound with excellent loudspeakers.

The result was astonishing. Many subjects reported seeing a much better picture in B. The picture quality was in fact the same. But the viewing experience was considerably better. We tend to judge our experiences as a sensory whole, not by the parts. This important observation is sometimes missed in the design of VR systems.

In the design of military tank trainers, considerable effort was made to have the highest achievable display quality (at almost any cost), so that looking at the display was as close to looking out the small window of a tank as possible. Fine. Only after painstaking endeavors to keep increasing the number of scan lines did the designers think to introduce an inexpensive motion platform that vibrated a little. By further including some additional sensory effects—tank motor and tread sounds—so much "realism" was achieved that the designers were then able to reduce the number of scan lines; they nonetheless exceeded the requirement that the system look and feel real.

I am constantly asked why I wear my reading glasses when I eat, because I obviously do not need glasses to see my food or fork. My answer is simply that the food tastes better when I wear glasses. Seeing the food clearly is part of a meal's quality. Looking and feeling add to each other.

LOOKING AND FEELING 10

LOOK AT YOU

Personal computers are less able to sense human presence than are modern toilets or outdoor floodlights that have simple motion sensors. Your inexpensive auto-focus camera has more intelligence about what is in front of it than any terminal or computer system.

When you lift your hands from your computer keyboard, it does not know whether the pause is reflective, a nature break, or an interruption for lunch. It cannot tell the difference between talking to you alone or in front of six other people. It does not know if you are in your night- or party clothes or no clothes at all. For all it knows, you could have your back to it while it was showing you something important, or you could be out of earshot altogether while it was speaking to you.

We think today solely from the perspective of what would make it easier for a person to use a computer. It may be time to ask what will make it easier for computers to deal with humans. For example, how can you possibly hold a conversation with people if you don't even know they are there? You can't see them, and you don't know how many there are. Are they smiling? Are they even paying attention? We talk longingly about human-computer interactions and conversational systems, and yet we are fully prepared to leave one participant in this dialogue totally in the dark. It is time to make computers see and hear.

Research on and the application of computer vision has been almost exclusively devoted to scene analysis, especially for military uses, such as autonomous vehicles and smart bombs. Applications in outer space are also compelling and drive the state of the art. If you have a robot roaming around the moon, it cannot just transmit the video of what it sees back to a human operator on earth, because it takes too long for the signal to travel, even at the speed of light. If the robot comes upon a precipice, by the time the human operator has seen the video image of the cliff and sent a message back to the moon telling the robot to stop moving forward, the robot by then will have fallen over the edge. This case is one in which the robot must depend on its own judgment based on what it sees.

Scientists have made steady progress in image understanding and have developed techniques, for example, to derive shape from shading or separate objects from backgrounds. Only recently have they started examining computer recognition of people to improve the human-computer interface. Your face is, in effect, your display device, and your computer should be able to read it, which requires the recognition of your face and its unique expressions.

Our grimaces are tightly wired to our expressive intent. When we talk on the telephone our facial expressions are not turned off just because the person at the other end cannot see them. In fact, we sometimes contort our faces and gesture even more, in order to give greater emphasis and prosody to our spoken language. By sensing facial expressions, the computer accesses a redundant, concurrent signal that can enrich both spoken and written messages.

The technical challenge of recognizing faces and facial expressions is formidable; nevertheless, its realization is eminently achievable in some contexts. In applications that involve you and your computer, it only needs to know if it is you, as opposed to anybody else on the planet. In addition, the backgrounds are easily separable.

Computers are likely to look at you sooner rather than later. During the 1990–91 Persian Gulf War when much corporate travel was banned, we witnessed a huge growth in teleconferencing. More and more, personal computers have since become equipped with low-cost teleconferencing hardware.

Teleconferencing hardware consists of a TV camera centered above the display and of the hardware or software necessary to encode, decode, and display real-time video and put it on part or all of the computer screen. Personal computers will increasingly become vision-ready. Teleconferencing system designers did not think of using the camera for the personal computer to enjoy face-to-face communication, but why not?

OF MICE AND MEN

Neil Gershenfeld of the Media Lab compares a $30 mouse, which takes a few minutes to learn to use, with a $30,000 cello bow, which requires a lifetime to master. He contrasts the sixteen bow techniques with the click, double-click, and drag of a mouse. The bow is for the virtuoso, and the mouse is for the rest of us.

A mouse is a simple but cumbersome medium for graphical input. It requires four steps: 1) groping with your hand to find the mouse, 2) wiggling the mouse to find the cursor, 3) moving the cursor to where you want it, 4) clicking or double-clicking the button. Apple's innovative design of the Power-Book at least reduces these steps to three and has the "dead mouse" (or more recently a track pad) where your thumbs are anyway, so typing interruptions are minimized.

Where mice and track balls really become useless is drawing. Try signing your name with a track ball. For such purposes a data tablet, a flat surface with a ballpoint-pen-like stylus, is a far better solution.

Not many computers have a data tablet for drawing. Those that do have the schizophrenic problem of properly situating both the tablet and keyboard, since each competes for centrality and wants to be directly in front of and below the display. The conflict is often resolved by putting the keyboard below the display because many people (including me) don't touch type.

As a result of both off-axis data tablets and mice off to the side, we must learn some rather unnatural hand-eye coordination. You draw or point in one place and look at another: touch drawing, so to speak.

Douglas Englebart, the man who invented the mouse in 1964, did so for pointing at text, not drawing. The invention stuck; we have them today everywhere. Jane Alexander, who chairs the National Endowment of the Arts, recently pointed out that only a man would have called it a mouse.

A year earlier, Ivan Sutherland had perfected the concept of a light pen for drawing directly on the screen (in the 1950s the SAGE defense system had some crude light pens). It tracked a cross-shaped cursor made up of five dots of light. To terminate drawing, he would flick his wrist and purposely lose tracking— a cute but not so accurate way to terminate a line.

Light pens are virtually nonexistent today. Holding your hand up to a screen is one thing (already hard to do for long, as the blood runs out of your hand), but carrying a tethered two-ounce pen causes extreme hand and arm fatigue. In some cases light pens were half an inch in diameter, and it felt like writing a postcard with a cigar.

Data tablets are particularly comfortable for drawing and, with some effort, the stylus can be made to have the texture and richness of an artist's paintbrush. To date, they have tended to be ballpoint pens (in feeling) on a flat and hard surface that needs desktop real estate somewhere close to you and your display. Since our desks are already so cluttered, the only way data tablets are likely to become popular is if furniture manufacturers start building them into desktops, so there is no device, as such, but the desktop itself.

HIGH-TOUCH COMPUTING

The dark horse in graphical input is the human finger.

Automated bank teller machines and information kiosks now use touch-sensitive displays quite successfully. But personal computers almost never have your finger meet the display, which is quite startling when you consider that the human finger is a pointing device you don't have to pick up, and we have ten of them. You can move gracefully from typing to pointing, from horizontal plane to vertical. Yet it hasn't caught on. Here are the three reasons I am given, but I do not, for a moment, believe any of them.

You cover what you point at when you point at it. True, but that happens with paper and pencil and has not stopped the practice of handwriting or using a finger to identify something on hard copy.

Your finger is low resolution. False. It may be stubby, but it has extraordinary resolution. It just takes a second step, after touching a surface, to roll your finger gently and to position a cursor with great accuracy.

Your finger dirties the screen. But it also cleans the screen! One way to think about touch-sensitive displays is that they will be in a steady state of more or less invisible filth, in which clean hands clean and dirty ones dirty (sort of).

The real reason for not using fingers is that we haven't yet found a good technology for sensing the near field of a finger: when your finger is close to but not touching the display. With just two states, touching or not touching, many applications are awkward at best. Whereas if a cursor appeared when your finger was within, say, a quarter of an inch of the display, then touching the screen itself could be like a mouse click.

A final feature of fingers is that the papillary ridges that make up your fingerprint also behave like treads on a snow tire and add friction where the skin meets the glass. This adhesion allows you to actually push on the screen and introduce forces into the plane of the screen.

In a device we built at MIT twenty years ago, we showed that when touching hard with your finger, without moving it, there was enough friction for you to put objects into motion, push and pull on them, and even induce rotational forces. One demonstration showed knobs on the screen that you could touch with two or three fingers and rotate owing to the adherence of your fingers to the display. The knobs not only turned but also made a clicking sound, which added even more realism. Possible applications include everything from children's games to the simplification of an airplane's cockpit.

THE INTERFACE STRIKES BACK

Remote manipulators have commonly been used in human-toxic environments like nuclear reactors. The robot arm would be inside the reactor and a human operator would be controlling it from the outside. Typically, the master and slave arms would be far apart, and the operator would be looking at a TV image of the scene. The slave end usually would have a pinching claw controlled at the master end by the operator's thumb and index finger, allowing objects to be picked up by grasping them; in this way a piece of uranium's weight and its elasticity (if any) can be felt.

Fred Brooks and his colleagues at the University of North Carolina had a marvelous idea: imagine that the slave arm did not exist at all, but the wires leading to it were instead connected to a computer that simulates the whole experience. The objects you see on the screen are now not real, but modeled and displayed by the computer with all the features of weight and elasticity.

The tactile properties of a computer have almost exclusively been thought of as you touching it, not vice versa.

I was involved in building an early prototype of a machine that pushed back at you, a force-feedback device in which the effort required to move it could be a function of anything you wanted. Under computer control, it could change from moving freely to feeling as if it were being pushed through molasses. In one application, we had a map of Massachusetts with a demographic database. The user could plot plans for a new highway by moving this force-feedback digitizer. However, the amount of force needed to push it varied as a function of the number of families that would be displaced. In fact, you could close your eyes and plot out very literally the path of least resistance to a new highway.

When IBM introduced the little red joystick (a mouse substitute) in the middle of their ThinkPad's keyboard, they opened themselves to this kind of force-feedback application (because it works by sensing force, not by sensing displacement). Hopefully the market will soon enjoy more widespread use of high-touch computing, when applications evolve that let the ThinkPad's joystick feel as if it is pushing back.

Another example was demonstrated at Apple Computer by Alan Kay (who is generally considered the father of personal computers). One of his researchers designed a "stubborn" mouse that used a variable magnetic field to make it more or less difficult to move. By adding full magnetic juice, the mouse would come to a total halt and move no further, keeping the cursor out of forbidden areas.

EYEING YOUR COMPUTER

Imagine reading a computer screen and being able to ask, What does that mean? Who is she? How did I get there? *That, she,* and *there* are defined by the direction of your gaze at the moment. Your questions concern the point of contact between your eyes and the text. Eyes are not normally considered output devices, yet we use them that way all the time.

The way humans can detect each other's direction of gaze and make eye contact is best described as magic. Think of standing twenty feet away from another person who at times is looking you right in the eye and at other times is looking just over your shoulder. You can instantly tell the difference even if that person's direction of gaze is just a fraction of a degree off-axis with your own. How?

It surely isn't trigonometry, where you are computing the angle of a normal to the plane of the other person's eyeballs and then computing whether that normal intersects with your own line of view. No. Something else is happening—a message is passing between your eyes and that person's. We do not have a clue as to how this works.

We use our eyes to point at objects all the time. When asked where somebody went, your answer may just be to look at an open door. When explaining what to carry, you may stare at one suitcase versus another. This kind of pointing, combined with head gesture, can be a very powerful channel of communication.

Several technologies exist to track eyes. One of the first demonstrations I ever saw was a head-mounted eye tracker that changed text on the screen from English into French as you read it. As your center of vision moved from word to word, you would see French words and the screen would seem to be 100 percent French. An onlooker, whose eyes were not being tracked, would see a screen full of roughly 99 percent English (namely, all the words except the one being looked at by the person wearing the eye tracker).

More modern eye-tracking systems deploy remote TV cameras so the user does not have to wear any device. A video-ready teleconferencing configuration is especially well suited for eye tracking, because the user tends to sit more or less in front of the screen at a relatively fixed distance. And often you will be looking into the eyes of the remote person (the computer can know where they are).

The more the computer knows about your position, posture, and the particular nature of your eyes, the easier it is for it to know where you are looking. Ironically, this seemingly exotic medium of eyes-as-input will find its first application in the rather commonplace configuration of a person seated at a desktop computer.

It will work even better when used concurrently with another channel of input—speech.

CAN WE TALK
ABOUT THIS?

11

BEYOND WORDS

For most people, typing is not an ideal interface. If we could speak to our computers, even the most confirmed Luddite would use them more enthusiastically. Nonetheless, computers remain more or less deaf and mute. Why?

The primary reason for so few advances in speech recognition is lack of perspective, not lack of technology. When I see speech-recognition demonstrations or product advertisements with people holding microphones to their mouths, I wonder, Have they really overlooked the fact that one of the major values of speech is that it leaves your hands free? When I see people with their faces up next to the screen—talking—I wonder, Have they forgotten that the ability to function from a distance is a reason to use voice? When I hear people claim or demand user-independent recognition, I ask myself, Have they forgotten that we are talking to personal, not shared, computers? Why does everybody seem to be addressing the wrong parts of the problem?

Simple. Until recently, we have been driven by two misguided obsessions. The first was driven by old-fashioned telephonic communications, which wanted anybody, anywhere, to be able to pick up a telephone handset and issue spoken commands to a computer instead of conversing with a human operator—it should not matter if you speak with a Southern drawl, with a Bahstan accent, or in New Yawkese. The other obsession came from office automation—the talking typewriter, to which we speak nonstop, and it transcribes perfectly. Focusing on these two goals alone delayed us for years from attaining some more achievable (and useful) objectives: recognizing and understanding conversational speech in a highly personalized and interactive environment.

We also overlooked the value of speech beyond words. For example, computers today demand your absolute and full attention. Usually, you must be seated. You must attend, more or less exclusively, to both the process and content of the interaction. There is almost no way to use a computer *in passing* or to have it engage in one of several conversations. Speech will change this.

Being able to use a computer *beyond arm's length* is also very important. Imagine if talking to a person required that his or her nose always be in your face. We commonly talk to people at a distance, we momentarily turn away and do something else, and it is not so rare to be out of sight while still talking. I want to have a computer be within "earshot," which requires solving the problem of segregating speech from other ambient sounds, like those of the air conditioner or an airplane overhead.

Speech is also more than words in that it has parallel *subcarriers* of information. Anybody who has a child or a pet knows that how you say something can be much more important than what you say. Tone of voice is very important. Dogs, for example, respond almost totally to tone of voice and have very little innate ability to do complex lexical analysis, in spite of the exaggerated claims of their adoring owners.

Spoken words carry a vast amount of information beyond the words themselves. While talking, one can convey passion, sarcasm, exasperation, equivocation, subservience, and exhaustion—all with the exact same words. In speech recognition by computers, these nuances have been ignored or, worse, treated as bugs rather than features. They are, however, the very qualities that make speaking a richer medium than typing.

THREE DIMENSIONS IN SPEECH RECOGNITION

If you speak a foreign language moderately well, but not perfectly, you will find that listening to radio news with static is very hard or impossible. By contrast, somebody fluent in that language will find the noisy signal annoying at most. Recognition and understanding are intimately intertwined.

Computers currently don't have understanding in the sense that you and I can agree that we *know* what something means. Whereas in the future computers will no doubt become more intelligent, for the time being we are forced to solve problems of machine recognition without very much machine understanding. Separating these two tasks provides a clear path for translating spoken words into computer-readable commands. The problem of speech recognition has three variables: vocabulary size, degree of speaker independence, and word connectedness, the extent to which words can be slurred together as they are in the cadence of normal human speech.

Think of these dimensions of speech recognition as three axes. On the vocabulary axis, the fewer words to recognize, the easier it is for the computer. If the system knows in advance who is speaking, the problem is simpler. And if words must be enunciated separately, it is also easier.

The origin of these axes is the place where we find the smallest vocabulary of totally speaker-dependent words that must be uttered with distinct—pauses—between—each.

As we add to or move up any axis, as when making the vocabulary larger, making the system work for any speaker, or allowing words to be run together, the problem gets harder and harder. In the extreme case, we expect the computer to recognize any word, spoken by anybody, "inneny" degree of connectedness. A common assumption has been that we must be at the far extreme on most or all of these axes for speech recognition to be of any use whatsoever. Nonsense!

Consider each in turn. When it comes to vocabulary size, one might ask, How big is big enough: five hundred, five thousand, or fifty thousand words? But the question should really be, How many recognizable words are needed in the computer's memory at any one time? This question suggests breaking vocabularies into contextual subsets, so that chunks can be folded into the machine as needed. When I ask my computer to place a phone call, my Rolodex is loaded. When I am planning a trip, the names of places are there instead.

If one views vocabulary size as the set of words needed at any one time (call them "word windows"), then the computer needs to select from a far less daunting number of utterances: closer to five hundred than fifty thousand.

The presumed need for speaker independence is a telephone company requirement of the past, when a central computer had to be able to understand anybody, a kind of "universal service." Today, our computation is more widely distributed and personalized. We can do more recognition at the periphery of the network: in the PC, in the handset, or with help from a smart card. If I want to talk with an airline's computer from a telephone booth, I could call my home computer or take out my pocket computer and let it do the translation from voice to a machine-readable signal.

The slurring and blurring of words is the third issue. We don't want to talk to a computer like a tourist addressing a foreign child, deliberately mouthing each word and pausing after it. This axis is the most challenging, but it can be partially simplified by looking at language as multiword utterances, not just as single words. In fact, handling runtogetherspeech in this fashion may well be part of the personalization and training of your computer.

We can work very close to the easiest corner of speech recognition when we consider speech as an interactive and conversational medium.

PARAVERBALS

Speech is a medium often filled with sounds that cannot be found in a dictionary. Speech is not only more colorful than black-and-white text, but it can gain added meaning from conversational features, like, er, paraverbals.

In 1978 at MIT we used an advanced speaker-dependent, connected speech-recognition system that, like all such systems then and now, was subject to error when the speaker's voice showed even the lowest level of stress. When graduate students demonstrated it to our sponsors, we wanted the system to work perfectly. This anxiousness invariably produced enough stress in the voice of the graduate student who was doing the demonstration to cause the system to crash.

A few years later, another student had a terrific idea: find the pauses in the user's speech and program the machine to generate the utterance "aha" at judicious times. Thus, as one spoke to the machine, it would periodically say, a-haaa, aaa-ha, or aha. This had such a comforting effect (it seemed that the machine was encouraging the user to converse), that the user relaxed a bit more, and the performance of the system skyrocketed.

This concept revealed two important points: first, not all utterances need have lexical meaning to be valuable in communications; second, some utterances are purely conversational protocols. When you are on the telephone, if you do not say "aha" to the caller at appropriate intervals, the person will become nervous and ultimately inquire, "Are you there?" The "aha" is not saying "yes," "no," or "maybe," but is basically transmitting one bit of information: "I'm here."

THE THEATER OF CONVERSATION

Imagine the following situation. You are sitting around a table where everyone but you is speaking French. Your French is limited to one miserable year in high school. A person at the table suddenly turns to you and says, *"Voulez-vous encore du vin?"* You understand perfectly. Subsequently, that same person changes the conversation to, say, politics in France. You will understand nothing unless you are fluent in French (and even then it is not certain).

You may think that "Would you like some more wine?" is baby talk, whereas politics requires sophisticated language skills. True. But that is not the important difference between the two conversations.

When the person asked you if you wanted more wine, he probably had his arm stretched toward the wine bottle and his eyes pointed at your empty wine glass. Namely, the signals you were decoding were parallel and redundant, not just acoustic. Furthermore, all the subjects and objects were in the same space and time. These tactics made it possible for you to understand.

Redundancy is, once again, good. The use of parallel channels (gesture, gaze, and speech) is the essence of human communications. Humans naturally gravitate to concurrent means of expression. If you have a modest command of Italian you will have a very hard time talking to Italians over the telephone. When you arrive at an Italian hotel and find no soap in the room, you won't use the telephone; you will go down to the concierge and use your best Berlitz to ask for soap. You may even make a few bathing gestures.

In a foreign land, one uses every means possible to transmit intentions and read all the signals to derive even minimal levels of understanding. Think of a computer as being in such a foreign land—ours.

MAKING COMPUTERS TALK GOOD

Speech can be produced by a computer in two ways: by replaying a previously recorded voice or by synthesizing the sounds from letters, syllables, or (most likely) phonemes. Each has its advantages and disadvantages. Speech production is like the problem of music: you can store the sound (like a CD does) and replay it or you can synthesize it to reproduce it from the notes (like a musician does).

Reciting previously stored speech returns the most "natural"-sounding oral and aural communication, particularly if the stored speech is a complete message. For this reason, most telephone messages are thus recorded. When you try to paste together smaller prerecorded chunks of sound or discrete words, the results are less satisfactory because the overall prosody is missing.

In the old days, people were hesitant to use very much prerecorded speech for human-computer interaction, because it consumed so much memory. Today that is less of a problem.

The real problem is the obvious one. In order for stored speech to work at all, you have to have recorded it previously. If you expect your computer to say things with proper names, for example, all those names need to have been previously stored. Stored voice does not work for random speech. For this reason, the second method, synthesizing, is used.

A speech synthesizer takes a stream of text (no different than this sentence) and follows certain rules to enunciate each word, one by one. Each language is different and varies in its difficulty to synthesize.

English is one of the hardest, because we write (right and rite) it in such an odd and seemingly illogical way (weigh and whey). Other languages, such as Turkish, are much easier. In fact, Turkish is very simple to synthesize because Atatürk moved that language from Arabic to Latin letters in 1929 and, in so doing, made a one-for-one correspondence between the sounds and the letters. You pronounce each letter: no silent letters or confusing diphthongs. Therefore, at the word level, Turkish is a dream come true for a computer speech synthesizer.

Even if the machine can enunciate every and any word, the problem does not stop there. It is very difficult to give a collection of synthesized words an overall rhythm and emphasis at the phrase or sentence level, which is important not only to sound good but also to include color, expression, and tone in accordance to both content and intent. Otherwise, the result is a monotonic voice that sounds like a drunken Swede.

We are now seeing (hearing) some systems that combine synthesis and storage. As with most things digital, the long-term solution will be to use both.

ALL THINGS LARGE AND SMALL

In the next millennium, we will find that we are talking as much or more with machines than we are with humans. What seems to trouble people most is their own self-consciousness about talking to inanimate objects. We are perfectly comfortable talking to dogs and canaries, but not doorknobs or lampposts (unless you are totally drunk). Wouldn't I feel stupid talking to a toaster? Probably no more so than you used to feel talking to an answering machine.

One thing that will make this ubiquity of speech move more rapidly today than in the past is miniaturization. Computers are getting smaller and smaller. You can expect to have on your wrist tomorrow what you have on your desk today, what filled a room yesterday.

Many users of desktop computers don't fully appreciate the enormous reduction in size over the past ten years, because there are certain dimensions, like the size of a keyboard, that are kept as constant as possible and others, like the size of the display, that seek to get bigger, not smaller. Therefore, the overall size of a desktop machine is no smaller than a fifteen-year-old Apple II.

If you have not used a modem for a long time, the change in its size is far more indicative of the real change that has happened. Less than fifteen years ago, a 1200-baud modem (costing about $1,000) was almost the size of a toaster lying on its side. A 9600-baud modem, at that time, was a rack-mounted cage. Today, a 19,200-baud modem is found on a smart card. Even in this credit-card format, much of the real estate is unused and is there purely for reasons of form factors (to fill the slot and to be big enough to hold and not to lose). The main reason for not putting something like a modem on the "head of a pin" is no longer technological; it is that we have trouble keeping track of heads of pins and misplace them easily.

Once you abandon the constraint of the natural spread of your fingers, which determines what makes for a comfortable keyboard, a computer's size is then driven more by the size of pockets, wallets, wristwatches, ballpoint pens, and the like. In these form factors, where a credit card is close to the smallest size desirable, a display is tiny and a graphical user interface makes very little sense.

Pen-based systems are likely to be viewed as an awkward interim means, too big and yet too small. The alternative of physical buttons is also an unacceptable solution. Look at your VCR or TV remote-control unit, and you probably have a good example of the limit of buttons, made for Pygmy hands and very young eyes.

For all these reasons, the trend of increasing miniaturization is bound to drive the improvement of speech production and recognition as the dominant human-computer interface with small objects. The actual speech recognition need not reside fully in each cuff link and watchband. Small devices can telecommunicate for assistance. The point is that being small begs for voice.

REACH OUT AND TOUCH SOMEONE

Many years ago, the head of research at Hallmark cards explained to me that the company's primary competitor was AT&T. "Reach out and touch someone" is about the transmission of emotion through voice. The voice channel carries not only the signal but all the attendant features that make it have the traits of being understanding, deliberate, compassionate, or forgiving. We say that somebody "sounds" honest, that an argument "sounds" fishy, or that something does not "sound" like so-and-so. Embedded in sound is information about feelings.

In the same way that we reach out to touch someone we will find ourselves using voice to project our desires to machines. Some people will behave like drill sergeants toward their computers, and others will be the voice of reason. Speech and delegation are tightly coupled. Will you be issuing orders to the Seven Dwarfs?

Possibly. The idea that twenty years from now you will be talking to a group of eight inch-high holographic assistants walking across your desk is not farfetched. What is certain is that voice will be your primary channel of communication between you and your interface agents.

LESS IS MORE

12

DIGITAL BUTLERS

In December 1980, Jerome Wiesner and I were the overnight and dinner guests of Nobutaka Shikanai at his lovely country house in the Hakone region of Japan, not far from Mount Fuji. We were so convinced that Mr. Shikanai's newspaper and TV media empire would benefit from being part of the inception of the Media Lab that Mr. Shikanai would be willing to help pay for building it. We further believed that Mr. Shikanai's personal interest in contemporary art would play right into our dream of blending technology with expression, of combining the invention with the creative use of new media.

Before dinner, we walked around Mr. Shikanai's famous outdoor art collection, which during the daytime doubles as the

Hakone Open Air Museum. At dinner with Mr. and Mrs. Shikanai, we were joined by Mr. Shikanai's private male secretary who, quite significantly, spoke perfect English, as the Shikanais spoke none at all. The conversation was started by Wiesner, who expressed great interest in the work by Alexander Calder and told about both MIT's and his own personal experience with that great artist. The secretary listened to the story and then translated it from beginning to end, with Mr. Shikanai listening attentively. At the end, Mr. Shikanai reflected, paused, and then looked up at us and emitted a shogun-size "Ohhhh."

The male secretary then translated: "Mr. Shikanai says that he too is very impressed with the work of Calder and Mr. Shikanai's most recent acquisitions were under the circumstances of . . ." Wait a minute. Where did all that come from?

This continued for most of the meal. Wiesner would say something, it would be translated in full, and the reply would be more or less an "Ohhhh," which was then translated into a lengthy explanation. I said to myself that night, if I really want to build a personal computer, it has to be as good as Mr. Shikanai's secretary. It has to be able to expand and contract signals as a function of knowing me and my environment so intimately that I literally can be redundant on most occasions.

The best metaphor I can conceive of for a human-computer interface is that of a well-trained English butler. The "agent" answers the phone, recognizes the callers, disturbs you when appropriate, and may even tell a white lie on your behalf. The same agent is well trained in timing, versed in finding the opportune moments, and respectful of idiosyncrasies. People who know the butler enjoy considerable advantage over a total stranger. That is just fine.

Such human agents are available to very few people. A more widely played role of similar sorts is that of an office secretary. If you have somebody who knows you well and shares much of your information, that person can act on your behalf very effectively. If your secretary falls ill, it would make no difference if the temporary agency could send you Albert Einstein. This issue is not about IQ. It is shared knowledge and the practice of using it in your best interests.

The idea of building this kind of functionality into a computer until recently was a dream so far out of reach that the concept was not taken seriously. This is changing rapidly. Enough people now believe that such "interface agents" are buildable. For this reason, this backwater interest in intelligent agents has become the most fashionable topic of research in human-computer interface design. It has become obvious that people want to delegate more functions and prefer to directly manipulate computers less.

The idea is to build computer surrogates that possess a body of knowledge both about something (a process, a field of interest, a way of doing) and about you in relation to that something (your taste, your inclinations, your acquaintances). Namely, the computer should have dual expertise, like a cook, gardener, and chauffeur using their skills to fit your tastes and needs in food, planting, and driving. When you delegate those tasks it does not mean you do not like to prepare food, grow plants, or drive cars. It means you have the option to do those things when you wish, because you want to, not because you have to.

Likewise with a computer. I really have no interest whatsoever in logging into a system, going through protocols, and figuring out your Internet address. I just want to get my message through to you. Similarly, I do not want to be required to read thousands of bulletin boards to be sure I am not missing something. I want my interface agent to do those things.

Digital butlers will be numerous, living both in the network and by your side, both in the center and at the periphery of your own organization (large or small).

I tell people about the intelligent pager that I have and love: how it delivers in full sentences of perfect English only timely and relevant information, how it behaves so intelligently. The way it works is that only one human being has its number and all messages go through that person, who knows where I am, what is important, and whom I know (and their agent). The intelligence is in the head end of the system, not at the periphery, not in the pager itself.

But you should have intelligence at the receiving end as well. I was recently visited by the CEO of a large corporation and his assistant, who wore the CEO's pager and fed him its prompts at the most opportune moments. The assistant's functions of tact, timing, and discretion will eventually be built into the pager.

PERSONAL FILTERS

Imagine an electronic newspaper delivered to your home as bits. Assume it is sent to a magical, paper-thin, flexible, waterproof, wireless, lightweight, bright display. The interface solution is likely to call upon mankind's years of experience with headlining and layout, typographic landmarks, images, and a host of techniques to assist browsing. Done well, this is likely to be a magnificent news medium. Done badly, it will be hell.

There is another way to look at a newspaper, and that is as an interface to news. Instead of reading what other people think is news and what other people justify as worthy of the space it takes, being digital will change the economic model of news selections, make your interests play a bigger role, and, in fact, use pieces from the cutting-room floor that did not make the cut on popular demand.

Imagine a future in which your interface agent can read every newswire and newspaper and catch every TV and radio broadcast on the planet, and then construct a personalized summary. This kind of newspaper is printed in an edition of one.

A newspaper is read very differently on Monday morning than it is on Sunday afternoon. At 7 a.m. on a workday, you browse a newspaper as a means of filtering the information and personalizing a common set of bits that were sent to hundreds of thousands of people. Most people tend to trash whole sections of newspapers without a glance, browse some of the rest, and read very little in detail.

What if a newspaper company were willing to put its entire staff at your beck and call for one edition? It would mix headline news with "less important" stories relating to acquaintances, people you will see tomorrow, and places you are about to go to or have just come from. It would report on companies you know. In fact, under these conditions, you might be willing to pay the Boston *Globe* a lot more for ten pages than for a hundred pages, if you could be confident that it was delivering you the right subset of information. You would consume every bit (so to speak). Call it *The Daily Me*.

On Sunday afternoon, however, we may wish to experience the news with much more serendipity, learning about things we never knew we were interested in, being challenged by a crossword puzzle, having a good laugh with Art Buchwald, and finding bargains in the ads. This is *The Daily Us*. The last thing you want on a rainy Sunday afternoon is a high-strung interface agent trying to remove the seemingly irrelevant material.

These are not two distinct states of being, black and white. We tend to move between them, and, depending on time available, time of day, and our mood, we will want lesser or greater degrees of personalization. Imagine a computer display of news stories with a knob that, like a volume control, allows you to crank personalization up or down. You could have many of these controls, including a slider that moves both literally and politically from left to right to modify stories about public affairs.

These controls change your window onto the news, both in terms of its size and its editorial tone. In the distant future, interface agents will read, listen to, and look at each story in its entirety. In the near future, the filtering process will happen by using headers, those bits about the bits.

DIGITAL SISTERS-IN-LAW

The fact that *TV Guide* has been known to make larger profits than all four networks combined suggests that the value of information about information can be greater than the value of the information itself. When we think of new information delivery, we tend to cramp our thoughts with concepts like "info grazing" and "channel surfing." These concepts just do not scale. With a thousand channels, if you surf from station to station, dwelling only three seconds per channel, it will take almost an hour to scan them all. A program would be over long before you could decide whether it is the most interesting.

When I want to go out to the movies, rather than read reviews, I ask my sister-in-law. We all have an equivalent who is both an expert on movies and an expert on us. What we need to build is a digital sister-in-law.

In fact, the concept of "agent" embodied in humans helping humans is often one where expertise is indeed mixed with knowledge of you. A good travel agent blends knowledge about hotels and restaurants with knowledge about you (which often is culled from what you thought about other hotels and restaurants). A real estate agent builds a model of you from a succession of houses that fit your taste with varying degrees of success. Now imagine a telephone-answering agent, a news agent, or an electronic-mail-managing agent. What they all have in common is the ability to model you.

It is not just a matter of completing a questionnaire or having a fixed profile. Interface agents must learn and develop over time, like human friends and assistants. Easily said, but not easily done. Only very recently have we started to get a handle on computer models that learn about people.

When I talk about interface agents, I am constantly asked, "Do you mean artificial intelligence?" The answer is clearly "yes." But the question carries implicit doubts raised by the false hopes and hyped promises of AI in the past. In addition, many people are still not comfortable with the idea that machines will be intelligent.

Alan Turing is generally considered the first person to seriously propose machine intelligence in his 1950 paper, "Computer Machinery and Intelligence." Later pioneers, such as Marvin Minsky, continued Turing's deep interest in pure AI. They ask themselves questions about recognizing context, understanding emotion, appreciating humor, and shifting from one set of metaphors to another. For example, What are the subsequent letters in a sequence that starts *O, T, T, F, F?*

I think AI may have suffered a turn for the worse around 1975 when computing resources started to achieve the kind of power that might be needed to solve intuitive problems and to exhibit intelligent behavior. What happened is that scientists suddenly opted for the very doable and marketable applications, like robotics and expert systems (i.e., stock trading and airline reservations), thereby leaving untouched the more profound and basic questions of intelligence and learning.

Minsky is quick to point out that even while today's computers can exhibit an uncanny grasp of airline reservations (a subject almost beyond logic), they absolutely cannot display the common sense exhibited by a three- or four-year-old child. They cannot tell the difference between a dog and a cat. Subjects like common sense are now moving off the back burner onto the center stage of scientific research, which is very important because an interface agent without common sense would be a pain in the neck.

By the way, the answer to the question raised above is *S, S.* The sequence is determined by the first letter of each word as you count: *one, two, three, four,* etc.

DECENTRALIZATION

A future interface agent is often seen as some centralized and omniscient machine of Orwellian character. A much more likely outcome is a collection of computer programs and personal appliances, each of which is pretty good at one thing and very good at communicating with the others. This image is fashioned after Minsky's *The Society of Mind* (1987), in which he proposes that intelligence is not found in some central processor but in the collective behavior of a large group of more special-purpose, highly interconnected machines.

This view runs against a set of prejudices that Mitchel Resnick, in his 1994 book, *Turtles, Termites, and Traffic Jams,* calls the "centralized mind-set." We are all strongly conditioned to attribute complex phenomena to some kind of controlling agency. We commonly assume, for example, that the frontmost bird in a V-shaped flock is the one in charge and the others are playing follow-the-leader. Not so. The orderly formation is the result of a highly responsive collection of processors behaving individually and following simple harmonious rules without a conductor. Resnick makes the point by creating situations in which people are surprised to find themselves part of such a process.

I recently experienced such a demonstration by Resnick in the Kresge Auditorium at MIT. The audience of roughly 1,200 people was asked to start clapping and try to clap in unison. Without the slightest lead from Resnick, within less than two seconds, the room was clapping a single beat. Try it yourself; even with much smaller groups the result can be startling. The surprise shown by participants brings home how little we understand or even recognize the emergence of coherence from the activity of independent agents.

This is not to say that your calendar agent will start planning meetings without consulting your travel agent. But every communication and decision need not go back to a central authority for permission, which might be a crummy way to manage an airline reservation system, but this method is viewed more and more as a viable way to manage organizations and governments. A highly intercommunicating decentralized structure shows far more resilience and likelihood of survival. It is certainly more sustainable and likely to evolve over time.

For a long time, decentralism was plausible as a concept but not possible as an implementation. The effect of fax machines on Tiananmen Square is an ironic example, because newly popular and decentralized tools were invoked precisely when the government was trying to reassert its élite and centralized control. The Internet provides a worldwide channel of communication that flies in the face of any censorship and thrives especially in places like Singapore, where freedom of the press is marginal and networking ubiquitous.

Interface agentry will become decentralized in the same way as information and organizations. Like an army commander sending a scout ahead or a sheriff sending out a posse, you will dispatch agents to collect information on your behalf. Agents will dispatch agents. The process multiplies. But remember the way this started: it started at the interface where you delegated your desires, versus diving into the World Wide Web itself.

This model of the future is distinctly different from a human-factors approach to interface design. The look and feel of the interface certainly count, but they play a minor role in comparison to intelligence. In fact, one of the most widely used interfaces will be the little tiny hole (or two) in plastic or metal, through which your voice accesses a small microphone.

It is also important to see the interface agent approach as very different from the current rage about the Internet and browsing it with Mosaic. The Internet hackers can surf that medium, explore enormous bodies of knowledge, and indulge in all kinds of new forms of socialization. This strikingly widespread phenomenon is not going to abate or go away, but it is only one kind of behavior, one more like direct manipulation than delegation.

Our interfaces will vary. Yours will be different from mine, based on our respective information predilections, entertainment habits, and social behavior—all drawn from the very large palette of digital life.

digital life

THE POST-
INFORMATION AGE

13

BEYOND DEMOGRAPHICS

The transition from an industrial age to a post-industrial or information age has been discussed so much and for so long that we may not have noticed that we are passing into a post-information age. The industrial age, very much an age of atoms, gave us the concept of mass production, with the economies that come from manufacturing with uniform and repetitious methods in any one given space and time. The information age, the age of computers, showed us the same economies of scale, but with less regard for space and time. The manufacturing of bits could happen anywhere, at any time, and, for example, move among the stock markets of New York, London, and Tokyo as if they were three adjacent machine tools.

In the information age, mass media got bigger and smaller at the same time. New forms of broadcast like CNN and *USA Today* reached larger audiences and made broadcast broader. Niche magazines, videocassette sales, and cable services were examples of narrowcasting, catering to small demographic groups. Mass media got bigger and smaller at the same time.

In the post-information age, we often have an audience the size of one. Everything is made to order, and information is extremely personalized. A widely held assumption is that individualization is the extrapolation of narrowcasting—you go from a large to a small to a smaller group, ultimately to the individual. By the time you have my address, my marital status, my age, my income, my car brand, my purchases, my drinking habits, and my taxes, you have me—a demographic unit of one.

This line of reasoning completely misses the fundamental difference between narrowcasting and being digital. In being digital I am *me,* not a statistical subset. *Me* includes information and events that have no demographic or statistical meaning. Where my mother-in-law lives, whom I had dinner with last night, and what time my flight departs for Richmond this afternoon have absolutely no correlation or statistical basis from which to derive suitable narrowcast services.

But that unique information about me determines news services I might want to receive about a small obscure town, a not so famous person, and (for today) the anticipated weather conditions in Virginia. Classic demographics do not scale down to the digital individual. Thinking of the post-information age as infinitesimal demographics or ultrafocused narrowcasting is about as personalized as Burger King's "Have It Your Way."

True personalization is now upon us. It's not just a matter of selecting relish over mustard once. The post-information age is about acquaintance over time: machines' understanding individuals with the same degree of subtlety (or more than) we can expect from other human beings, including idiosyncrasies (like always wearing a blue-striped shirt) and totally random events, good and bad, in the unfolding narrative of our lives.

For example, having heard from the liquor store's agent, a machine could call to your attention a sale on a particular Chardonnay or beer that it knows the guests you have coming to dinner tomorrow night liked last time. It could remind you to drop the car off at a garage near where you are going, because the car told it that it needs new tires. It could clip a review of a new restaurant because you are going to that city in ten days, and in the past you seemed to agree with that reviewer. All of these are based on a model of you as an individual, not as part of a group who might buy a certain brand of soapsuds or toothpaste.

PLACE WITHOUT SPACE

In the same ways that hypertext removes the limitations of the printed page, the post-information age will remove the limitations of geography. Digital living will include less and less dependence upon being in a specific place at a specific time, and the transmission of place itself will start to become possible.

If I really could look out the electronic window of my living room in Boston and see the Alps, hear the cowbells, and smell the (digital) manure in summer, in a way I am very much in Switzerland. If instead of going to work by driving my atoms into town, I log into my office and do my work electronically, exactly where is my workplace?

In the future, we will have the telecommunications and virtual reality technologies for a doctor in Houston to perform a delicate operation on a patient in Alaska. In the nearer term, however, a brain surgeon will need to be in the same operating theater at the same time as the brain; many activities, like those of so-called knowledge workers, are not as dependent on time and place and will be decoupled from geography much sooner.

Today, writers and money managers find it practicable and far more appealing to be in the Caribbean or South Pacific while preparing their manuscripts or managing their funds. However, in some countries, like Japan, it will take longer to move away from space and time dependence, because the native culture fights the trend. (For example: one of the main reasons that Japan does not move to daylight savings time in the summer is because going home "after dark" is considered necessary, and workers try not to arrive after or go home before their bosses.)

In the post-information age, since you may live and work at one or many locations, the concept of an "address" now takes on new meaning.

When you have an account with America Online, Compu-Serve, or Prodigy, you know your own e-mail address, but you do not know where it physically exists. In the case of America Online, your Internet address is your ID followed by @aol.com—usable anywhere in the world. Not only do you not know where @aol.com is, whosoever sends a message to that address has no idea of where either it or you might be. The address becomes much more like a Social Security number than a street coordinate. It is a virtual address.

In my case, I happen to know where my address, @hq.media.mit.edu, is physically located. It is a ten-year-old HP Unix machine in a closet near my office. But when people send me messages they are sending them to me, not to that closet. They might infer I am in Boston (which is usually not the case). In fact, I am usually in a different time zone, so not only space but time is shifted as well.

BEING ASYNCHRONOUS

A face-to-face or telephone conversation is real time and synchronous. Telephone tag is a game played to find the opportunity to be synchronous. Ironically, this is often done for exchanges, which themselves require no synchrony whatsoever, and could just as well be handled by non-real-time message passing. Historically, asynchronous communication, like letter writing, has tended to be more formal and less off-the-cuff exchanges. This is changing with voice mail and answering machines.

I have met people who claim they cannot understand how they (and we all) lived without answering machines at home and voice mail at the office. The advantage is less about voice and more about off-line processing and time shifting. It is about leaving messages versus engaging somebody needlessly in on-line discussion. In fact, answering machines are designed slightly backward. They should not only activate when you are not there or don't want to be there, but they should *always* answer the telephone and give the caller the opportunity to simply leave a message.

One of the enormous attractions of e-mail is that it is not interruptive like a telephone. You can process it at your leisure, and for this reason you may reply to messages that would not stand a chance in hell of getting through the secretarial defenses of corporate, telephonic life.

E-mail is exploding in popularity because it is *both* an asynchronous and a computer-readable medium. The latter is particularly important, because interface agents will use those bits to prioritize and deliver messages differently. Who sent the message and what it is about could determine the order in which you see it—no different from the current secretarial screening that allows a call from your six-year-old daughter to go right through, while the CEO of the XYZ Corporation is put on hold. Even on a busy workday, personal e-mail messages might drift to the top of the heap.

Not nearly as much of our communications need to be contemporaneous or in real time. We are constantly interrupted or forced into being punctual for things that truly do not merit such immediacy or promptness. We are forced into regular rhythms, not because we finished eating at 8:59 p.m., but because the TV program is about to start in one minute. Our great-grandchildren will understand our going to the theater at a given hour to benefit from the collective presence of human actors, but they will not understand the synchronous experiencing of television signals in the privacy of our home—until they look at the bizarre economic model behind it.

DEMANDING ON DEMAND

Digital life will include very little real-time broadcast. As broadcast becomes digital, the bits are not only easily time-shiftable but need not be received in the same order or at the same rate as they will be consumed. For example, it will be possible to deliver one hour of video over fiber in a fraction of a second (some experiments today show that the time needed to deliver one hour of VHS-quality video can be as small as one-hundredth of a second). Alternately, over a thin wire or narrow radio frequency, you might use six hours of broadcast time overnight to transmit a ten-minute (personalized) video news program. The former is blasting the bits into your computer and the latter is trickle-charging it.

With the possible exception of sports and elections, technology suggests that TV and radio of the future will be delivered asynchronously. This will happen either on demand or using "broadcatching," a term coined in 1987 by Stewart Brand in his book about the Media Lab. Broadcatching is the radiation of a bit stream, most likely one with vast amounts of information pushed into the ether or down a fiber. At the receiving end, a computer catches the bits, examines them, and discards all but the few it thinks you want to consume later.

On-demand information will dominate digital life. We will ask explicitly and implicitly for what we want, when we want it. This will require a radical rethinking of advertiser-supported programming.

In 1983, when we started the Media Lab, people felt that the word *media* was pejorative, representing a one-way path to the lowest common denominator in American culture. Media, with a capital *M,* almost exclusively meant "mass media." A large audience would bring in large advertising bucks, which in turn would underwrite large production budgets. Advertising was further justified in "over-the-air" mass media on the basis that information and entertainment should be "free" to the viewer, since spectrum is public property.

Magazines, on the other hand, use a private network of distribution and share the cost between advertiser and reader. Magazines, a notably asynchronous medium, offer a much wider range of economic and demographic models and may in fact be a bellwether for the future of television. The proliferation into niche markets has not necessarily ruptured content, but it has shifted some of the burden of cost to the subscriber. In some specialty magazines there is no advertising at all.

In future digital media there will be more pay-per-view, not just on an all-or-nothing basis, but more like newspapers and magazines, where you share the cost with advertisers. In some cases, the consumer may have an option to receive material without advertising but at a higher cost. In other cases, the advertising will be so personalized that it is indistinguishable from news. It *is* news.

The economic models of media today are based almost exclusively on "pushing" the information and entertainment out into the public. Tomorrow's will have as much or more to do with "pulling," where you and I reach into the network and check out something the way we do in a library or video-rental store today. This can happen explicitly by asking or implicitly by an agent asking on your behalf.

The on-demand model without advertising will make content production more like theatrical Hollywood movies, with much higher risk but much higher reward as well. There will be big busts and wild successes. Make it, and they will come. If they do, great; if they don't, too bad, but Procter & Gamble is not necessarily underwriting the risk. In this sense, media companies will be throwing bigger dice tomorrow than they do today. But there will also be smaller players, throwing smaller dice, getting part of the audience share.

The "prime" of prime time will be its quality in our eyes, not those of some collective and demographic mass of potential buyers of a new luxury car or dishwashing detergent.

PRIME TIME IS MY TIME

14

BITS FOR RENT (INQUIRE WITHIN . . .)

Many people believe that video-on-demand, VOD, will be the killer application to finance the superhighway. The reasoning goes like this: say a videocassette-rental store has a selection of four thousand tapes. Suppose it finds that 5 percent of those tapes result in 60 percent of all rentals. Most likely, a good portion of that 5 percent would be new releases and would represent an even larger proportion of the store's rentals if the available number of copies were larger.

After studying these videocassette-rental habits, the easy conclusion is that the way to build an electronic VOD system is to offer only those top 5 percent, primarily new releases. Not only would this be convenient; it would provide tangible and convincing evidence for what some still consider an experiment.

Otherwise, it would take too much time and money to digitize many or all of the movies made in America by 1990. It would take even more time to digitize the quarter of a million films in the Library of Congress, and I'm not even considering the movies made in Europe, the tens of thousands from India, or the twelve thousand hours per year of *telenovelas* made in Mexico by Televisa. The question remains: Do most of us really want to see just the top 5 percent, or is this herd phenomenon driven by the old technologies of distributing atoms?

Blockbuster opened six hundred new stores in 1994 (occupying 5 million square feet) on the force of its entrepreneurial founding and former chairman, H. Wayne Huizenga, claiming that 87 million American homes took fifteen years to have a $30 billion investment in VCRs and that Hollywood has such a big stake in selling him cassettes that it would not dare enter into VOD agreements.

I don't know about you, but I would throw away my VCR tomorrow for a better scheme. The issue to me is one of schlepping (and returning) atoms (by what is sometimes called "sneaker net"), versus receiving no-return, no-deposit bits. With all due respect to Blockbuster and its new owner, Viacom, I think videocassette-rental stores will go out of business in less than ten years.

Huizenga has argued that pay-per-view television hasn't worked, so why should on-demand TV work? But videocassette-rentals *are* pay-per-view. In fact, the very success of Blockbuster proves that pay-per-view works. The only difference for the time being is that his stores, which rent atoms, are easier to browse than a menu of rentable bits. But this is changing rapidly. When electronic browsing is made more pleasant by imaginative agent-based systems, then, unlike Blockbuster, VOD will not be limited to a few thousand selections, but will be literally unlimited.

ANYTHING, ANYTIME, ANYWHERE TELEVISION

Some of the world's most senior telephone executives recite the jingle "Anything, anytime, anywhere" like a poem for modern mobility. But my goal (and I suspect yours) is to have "Nothing, never, nowhere" unless it is timely, important, amusing, relevant, or capable of reaching my imagination. AAA stinks as a paradigm for telecommunications. But it is a beautiful way to think about television.

When we hear about a thousand channels of TV, we tend to forget that, even without satellite, we already have more than a thousand programs per day delivered to our homes. Admittedly, they are sent at all—and odd—hours. When the 150-plus channels of TV listed in *Satellite TV Week* are included, the result is an additional 2,700 or more programs available in a single day.

If your TV could record every program transmitted, you would already have five times the selectivity offered in the superhighway's broad-brush style of thinking. Say, instead of keeping them all, you have your TV agent grab the one or two in which you might have interest, for your future viewing at any time.

Now let AAA-TV expand to a global infrastructure of fifteen thousand channels of television and the quantitative and the qualitative changes become very interesting. Some Americans might watch Spanish television to perfect their Spanish, others might follow Channel 11 on Swiss cable to see unedited German nudity (at 5 p.m. New York time), and the 2 million Greek Americans might find it interesting to see any one of the three national or seven regional channels from Greece.

More interesting, perhaps, is the fact that the British devote seventy-five hours per year to the coverage of chess championships and the French commit eighty hours of viewing to the Tour de France. Surely American chess and bicycle enthusiasts would enjoy access to these events—anytime, anywhere.

COTTAGE TELEVISION

If I were contemplating a visit to the southwestern coast of Turkey, I might not find a documentary on Bodrum, but I could find sections from movies about wooden-ship building, night-time fishing, underwater antiquities, baba ghanouj, and Oriental carpets from such sources as *National Geographic,* PBS, the BBC, and hundreds of others. These pieces could be woven together to form a story that would suit my specific need. The result would not likely win an Oscar for best documentary, but that's not the point.

VOD can provide a new life for documentary films, even the dreaded infomercial. Digital TV agents will edit movies on the fly, much like a professor assembling an anthology using chapters from different books and articles from different magazines. Copyright lawyers, fasten your seat belts.

On the Net each person can be an unlicensed TV station. Three and a half million camcorders were sold in the United States during 1993. Every home movie won't be a prime-time experience (thank God). But we can now think of mass media as a great deal more than high production value, professional TV.

Telecommunications executives understand the need for broadband into the home. They cannot fathom the need for a channel of similar capacity in the reverse direction. This asymmetry is justified by experience with interactive computer services that are sometimes offered with higher bandwidth going to you and lower bandwidth coming back from you. This is because, for example, most of us type much more slowly than we read, and recognize images much faster than we can draw them.

This asymmetry does not exist with video services. The channel needs to be two-way. An obvious example is teleconferencing, which will become a particularly valuable consumer medium for grandparents or, in divorced families, for the parent who does not have custody of the children.

That's live video. Consider "dead" video. In the near future, individuals will be able to run electronic video services in the same way that fifty-seven thousand Americans run computer bulletin boards today. That's a television landscape of the future that is starting to look like the Internet, populated by small information producers. In a few years you can learn how to make couscous from Julia Child or a Moroccan housewife. You can discover wines with Robert Parker or a Burgundian vintner.

THE TOPOLOGY OF A SHRINKING PLANET

There are currently four electronic paths into the home: telephone, cable, satellite, and terrestrial broadcast. Their differences are more about topology than alternate economic models. If I want to deliver the same bit at the same time to every household in the continental United States, I should obviously use a single satellite whose footprint reaches from East to West Coast. This would be the most logical topology, versus, for example, sending that bit to every single one of the twenty-two thousand telephone exchanges in the United States.

By contrast, if I have regional news or advertising, terrestrial broadcast works well, and cable even better. Telephone works best for point to point. If I decide which medium to use based solely on topology, I would put the Super Bowl on satellite and an interactive, personalized version of "Wall Street Week" over the telephone network. The delivery path—satellite, terrestrial broadcast, cable, or telephone—can be decided in terms of which is best suited for which kind of bit.

But in the "real world," as people are fond of telling me (as if I live in an unreal world), each channel tries to increase its payload, often by using itself to do what it does least well.

For example, some operators of stationary satellites are considering land-based point-to-point network services. This really makes little sense, in comparison to a wired telephone network, unless you are in a place that is trying to overcome some special geographic or political obstacle, like an archipelago or censorship. Similarly, to send the Super Bowl down every terrestrial, cable, or telephone system is a hard way to get those bits to everybody at the same time.

Slowly but surely, bits will migrate to the proper channel at the proper time. If I want to see last year's Super Bowl, calling it up by telephone is the logical way to accomplish this (versus waiting for somebody to rebroadcast it). After the game, the Super Bowl is suddenly archival data and the suitable channel is very different from when it was "live."

Each delivery channel has its own anomalies. When delivering a message by satellite from New York to London, the distance traveled by the signal is only five miles longer than from New York to Newark by the same method. This suggests that maybe a telephone call within the footprint of a given satellite should cost the same, whether you are calling from Madison to Park Avenue or from Times Square to Picadilly Circus.

Fiber will force similar reconsideration of pricing bit delivery. When a single trunk carries bits between New York and Los Angeles, it is unclear if conveying them that long distance is more or less costly than shipping through the highly switched capillary system of a suburban telephone network.

Distance means less and less in the digital world. In fact, an Internet user is utterly oblivious to it. On the Internet, distance often seems to function in reverse. I frequently get faster replies from distant places than close ones because the time change allows people to answer while I sleep—so it feels closer.

When a delivery system that looks more like the Internet is used in the general world of entertainment, the planet becomes a single media machine. Homes equipped today with movable satellite dishes already get a taste for the wide range of programming, with no geopolitical boundaries. The problem is how to cope with it.

SIGNALS WITH A SENSE
OF THEMSELVES

The best way to deal with a massive amount of available television is not to deal with it at all. Let an agent do that.

Although future computing machines will be as capable of understanding video narrative as you or me, for the next thirty years or so, machine understanding of video content will be limited to very specific domains, like face recognition at ATM machines. This is a far cry from having a computer understand from the video that Seinfeld has just lost another girlfriend. Therefore, we need those bits that describe the narrative with key words, data about the content, and forward and backward references.

In the next few decades, bits that describe the other bits, tables of contents, indexes, and summaries will proliferate in digital broadcasting. These will be inserted by humans aided by machines, at the time of release (like closed captions today) or later (by viewers and commentators). The result will be a bit stream with so much header information that your computer really can help you deal with the massive amount of content.

My VCR of the future will say to me when I come home, "Nicholas, I looked at five thousand hours of television while you were out and recorded six segments for you which total forty minutes. Your high school classmate was on the 'Today' show, there was a documentary on the Dodecanese Islands, etc. . . ." It will do this by looking at the headers.

These same header bits work very well for advertising, too. If you are in the market for a new car, you can have nothing but car ads on your screen this week. Furthermore, the car companies can embed local, regional, and national information in the headers so that your neighborhood dealer's clearance sale is dutifully included. This can be expanded into an entire shopping channel, which, unlike QVC, sells only things you really want, instead of zirconium rings.

The bits about the bits change broadcasting totally. They give you a handle by which to grab what interests you and provide the network with a means to ship them into any nook or cranny that wants them. The networks will finally learn what networking is about.

NETWORKS AND NETWORKS

Television networks and computer networks are almost the opposite of each other. A television network is a distribution hierarchy with a source (where the signal comes from) and many homogeneous sinks (where the signals go to).

Computer networks, on the other hand, are a lattice of heterogeneous processors, any one of which can act both as source and sink. The two are so totally different from each other that their designers don't even speak the same language. The rationale of the one is about as logical to the other as Islamic fundamentalism is to an Italian Catholic.

For example, when you send e-mail over the Internet the message is decomposed into packets and given headers with an address, and pieces are sent over a variety of different paths, through a variety of intermediate processors, which strip off and add other header information and then, quite magically, reorder and assemble the message at the other end. The reason that this works at all is that each packet has those bits-about-bits and each processor has the means to pull out information about the message from the message itself.

When video engineers approached digital television they took no lessons from computer network design. They ignored the flexibility of heterogeneous systems and information-packed headers. Instead, they argued among themselves about resolution, frame rate, aspect ratio, and interlace, rather than let those be variables. Broadcast TV doctrine has all the dogma of the analog world and is almost devoid of digital principles, like open architecture, scalability, and interoperability. This will change, but change has so far been very slow in coming.

The agent of change will be the Internet, both literally and as a model or metaphor. The Internet is interesting not only as a massive and pervasive global network but also as an example of something that has evolved with no apparent designer in charge, keeping its shape very much like the formation of a flock of ducks. Nobody is the boss, and all the pieces are so far scaling admirably.

Nobody knows how many people use the Internet, because, first of all, it is a network of networks. As of October 1994, more than forty-five thousand networks were part of the Internet. There were more than 4 million host processors (growing at more than 20 percent per quarter), but that is not a helpful measure for estimating the number of users. All that needs to happen is that one of those machines serves as a public gateway to, say, France's Minitel system, and all of a sudden you have an additional 8 million potential users on the Internet.

The state of Maryland offers the Internet to all of its residents, as does the city of Bologna, Italy. Obviously all these people don't use it, but in 1994, 20 million to 30 million people seemed to. My guess is that 1 billion people will be connected by the year 2000. This is based in part on the fact that the fastest-growing number of Internet hosts (percent change) in the third quarter of 1994 were Argentina, Iran, Peru, Egypt, the Philippines, the Russian Federation, Slovenia, and Indonesia (in that order). All showed more than 100 percent growth in that three-month period. The Internet, affectionately called the Net, is not North American any more. Thirty-five percent of the hosts are in the rest of the world, and that is the fast-growing part.

Even though I use the Internet every day of the year, people like me are considered wimps on the Net. I use it strictly for e-mail. More savvy users and those who have the time cruise around the Net like walking into and out of shops in a mall. You can literally move from machine to machine and do window-shopping using tools like Mosaic or riding bareback (closer to the metal). You can also join real-time discussion groups, so-called MUDs, a term coined in 1979 meaning "multi-user dungeons" (some people are embarrassed by the name and will claim that it means multi-user domains). A newer form of a MUD is a MOO (MUD object-oriented). In a very real sense, MUDs and MOOs are a "third" place, not home and not work. Some people today spend eight hours a day there.

In the year 2000 more people will be entertaining themselves on the Internet than by looking at what we call the networks today. The Internet will evolve beyond MUDs and MOOs (which sounds a bit too much like Woodstock in the 1960s here in the 1990s in digital form) and start to serve up a broader range of entertainment.

Internet Radio is certainly a bellwether for the future. But even Internet Radio is the tip of the iceberg, because it is, so far, not much more than narrowcasting to a special kind of computer hacker, as evidenced by one of its major talk shows, called "Geek of the Week."

The user community of the Internet will be in the mainstream of everyday life. Its demographics will look more and more like the demographics of the world itself. As both Minitel in France and Prodigy in the United States have learned, the single biggest application of networks is e-mail. The true value of a network is less about information and more about community. The information superhighway is more than a short cut to every book in the Library of Congress. It is creating a totally new, global social fabric.

GOOD CONNECTIONS · 15

BEING DIGITAL IS NOT ENOUGH

When you read this page, your eyes and brain are converting the print medium into signals you can process and recognize as letters and words with meaning. If you were to fax this page, the scanner in the facsimile machine would generate a fine line-by-line map with 1s and 0s representing the black and white of ink and no ink. The faithfulness of the digital image to the actual page will vary in accordance with how finely it is scanned. But no matter how finely your fax scans the text, in the end your fax is nothing more or less than a picture of the page. It is not letters and words; it is pixels.

For the computer to interpret any of the content of that image, it must go through a process of recognition similar to yours. It must convert small areas of pixels into letters, and then those into words. This includes all the concomitant problems of distinguishing between the letter O and digit 0, separating a doodle from text, telling the difference between a coffee stain and an illustration, and discerning all of this on a background of noise, the freckles introduced by the scanning or transmission process.

Once this is done, your digital representation is no longer an image, but structured data in the form of letters, typically encoded in the binary representation called ASCII (which stands for American Standard Code for Information Interchange), plus some additional data about the Berkeley typeface and its layout on this page. This fundamental difference between fax and ASCII extends to other media.

A CD is "audio fax." It's digital data that allow us to compress, error-correct, and control the acoustic signal, but it does not provide the musical structure. It would be very hard, for example, to subtract the piano, to substitute a singer, or to change the spatial location of instruments in the orchestra. The dramatic difference between audio fax and a more structured representation of music was observed eight years ago by Mike Hawley, then a student and now a new faculty member at MIT, as well as a gifted pianist.

Hawley's Ph.D. studies included work with a specially instrumented Bosendorfer grand piano, which records the onset time of each hammer and the velocity with which it hits the wire. In addition, each of the keys is motorized so that it can play back with near-perfect performance. Think of it as a very fine-grained keyboard digitizer combined with the world's most expensive and high-resolution player piano. Recently, Yamaha has introduced a low-cost version of the same.

Hawley considered the problem of how to store more than one hour of music on a CD. The problem is being tackled by industry in two very incremental ways. One is to change the laser from red to blue, thereby shortening the wavelength and increasing the density by a factor of four. Another is to use more contemporary encoding techniques, because your CD player really is using algorithms from the mid-1970s and, since that time, we have learned how to compress audio better by at least a factor of four (with the same degree of so-called losslessness). Use these two techniques together, and you get a whopping sixteen hours of audio on one side of a CD.

Hawley explained to me one day that he had a way of putting many more hours of audio on a CD. When I asked, "How many?" he said, "About five thousand hours." Should this be true, I thought, the Music Publishers Association of the World would put a contract out on Hawley's life, and he would need to live with Salman Rushdie forever after. But I asked him to explain anyway (and we would keep it a secret, I said, with fingers crossed).

What Hawley had noticed with the Bosendorfer, and through his using it together with such household names as John Williams, is that human hands, playing very fast movements, just could not generate more than 30,000 bits per minute of output from the Bosendorfer. Namely, the gestural data, the measurement of finger movement, was very low. Compare this to the 1.2 million bps generated by audio, as on a CD. Namely, if you store the gestural, not the audio, data, you can indeed store five thousand times as much music. And you would not need a $125,000 Bosendorfer, but could use a more modest instrument equipped with MIDI (Musical Instrument Data Interface).

Everybody in industry who has looked at the problem of capacity in audio CDs has understandably but sheepishly addressed the problem as being exclusively in the audio domain, much like fax is in the image domain. By contrast, Hawley's insight is that gesture is like MIDI, and both are closer to ASCII. In fact, the musical score itself is an even more compact representation (admittedly low resolution and missing the expressive nuance that comes from human interpretation).

By looking for the structure in signals, how they were generated, we go beyond the surface appearance of bits and discover the building blocks out of which the image, sound, or text came. This is one of the most important facts of digital life.

THE FAX OF LIFE

Twenty-five years ago, if the computer science community had predicted the percentage of new text that would be computer-readable today, they would have estimated it to be as high as 80 or 90 percent. Until around 1980 they could have been right. Enter the facsimile machine.

The fax machine is a serious blemish on the information landscape, a step backward, whose ramifications will be felt for a long time. This condemnation appears to fly in the face of a telecommunications medium that has seemingly revolutionized the way we conduct business and, increasingly, our personal lives. But people don't understand the long-term cost, the short-term failings, and the alternatives.

Fax is a Japanese legacy, but not just because they were smart enough to standardize and manufacture them better than anybody else, like VCRs. It is because their culture, language, and business customs are very image-oriented.

As recently as ten years ago, Japanese business was not conducted via documents, but by voice, usually face to face. Few businessmen had secretaries, and correspondence was often painstakingly handwritten. The equivalent of a typewriter looked more like a typesetting machine, with an electro-mechanical arm positioned over a dense template of choices to produce a single Kanji symbol out of a total of more than sixty thousand.

The pictographic nature of Kanji made the fax a natural. Since little Japanese was then in computer-readable form, there were few disadvantages. On the other hand, for a language as symbolic as English, fax is nothing less than a disaster as far as computer readability is concerned.

With the twenty-six letters of the Latin alphabet, ten digits, and a handful of special characters, it is much more natural for us to think in terms of 8-bit ASCII. But the fax has made us ignore this. For example, most business letters today are prepared on a word processor, printed out, and faxed. Think about that. We prepare our document in completely computer-readable form, so readable in fact that we think nothing about passing a spell-checker over the words.

Then what do we do? We print it on paper bond letterhead. The document has now lost all of the properties of being digital.

We next take this piece of paper over to a fax machine, where it is (re)digitized into an image, removing what little qualities of feel, color, and letterhead that might have been in the paper. It is dispatched to a destination, perhaps a wire basket next to the copying machines. If you are one of the less fortunate recipients, you then get to read it on cancerous-feeling, slimy paper, sometimes uncut and reminiscent of ancient scrolls. Give me a break. This is about as sensible as sending each other tea leaves.

Even if your computer has a fax modem, which avoids the intermediary paper step, or even if your fax is plain paper and full color, fax is still not an intelligent medium. The reason is that you have removed the computer readability, which is the means by which the recipient can automatically store, retrieve, and manipulate your message.

How many times have you remembered a fax as having been received about six months ago from somebody . . . someplace . . . in reference to such-and-such? In ASCII form, one need only search a computer database for the occurrence of "such-and-such."

When you fax a spreadsheet all you can send is an image of it. With e-mail you can send an executable spreadsheet for the recipient to manipulate and query or to see in whatever form he or she desires.

Fax is not even economical. This page takes about twenty seconds to send by normal fax at 9600 baud. This represents approximately 200,000 bits of information in that form. On the other hand, using electronic mail, less than one-tenth of those bits are necessary: the ASCII and some control characters. In other words, even if you claim not to care about computer readability, e-mail is 10 percent of the cost of fax, measured per bit or per second running at the same 9600 baud (at 38,400 baud it is 2.5 percent the cost of a current fax).

The idea of both fax and electronic mail goes back about a hundred years. In an 1863 manuscript, "Paris in the 20th Century," found and published for the first time in 1994, Jules Verne wrote, "Photo-telegraphy allowed any writing, signature or illustration to be sent far away, and any contract to be signed at a distance of [20,000 km]. Every house was wired."

Western Union's automatic telegraph (1883) was hard-wired, point-to-point e-mail. The general use of e-mail as we know it today, multipoint to multipoint, predates the general use of fax. When e-mail started during the middle and late 1960s, relatively few people were computer literate. Therefore, it is not surprising that e-mail was dramatically overtaken by fax in the 1980s. The reasons were the ease of use, simple delivery of images and graphics, and input from hard copy (including forms). Also, under certain conditions and as of very recently, faxes have legal value with signatures.

But today, with computer ubiquity, the advantages of e-mail are overwhelming, as evidenced by its skyrocketing use. Beyond the digital benefits, e-mail is a more conversational medium. While it is not spoken dialogue, it is much closer to speaking than writing.

I always look at my e-mail first thing in the morning, and later in the day I am capable of saying, "Yes, I spoke to so-and-so this morning," whereas it was only e-mail. Messages are bounced back and forth. Such exchanges are often with typographical errors. I remember apologizing to a Japanese colleague for my typos, to which he replied that I should not worry because he was a far better spelling corrector than any software package I could buy. True indeed.

This new quasi-conversational medium is really very different from letter writing. It is far more than a fast post office. Over time, people will find different styles of usage. There is already a whole e-mail language for tone, using things like :)—a smiley face. In all likelihood, in the next millennium e-mail (by no means limited to ASCII) will be the dominant interpersonal telecommunications medium, approaching if not overshadowing voice within the next fifteen years. We will all be using e-mail, provided we all learn some digital decorum.

NETIQUETTE

Imagine the following scene: the ballroom of an Austrian castle during the eighteenth century, in full splendor, glittering with gilt and reflected light from hundreds of candles, Venetian mirrors, and jewels. Four hundred handsome and beautiful people waltz gracefully to a ten-piece orchestra—just like the ballroom scenes from Paramount's *The Scarlet Empress* or Universal Pictures' *The Merry Widow*. Now imagine the same scene, but 390 of the guests learned how to dance the night before; they are all too conscious of their feet. This is similar to the Internet today: most users are all thumbs.

The vast majority of Internet users today are newcomers. Most have been on it for less than a year. Their first messages tend to flood a small group of select recipients, not only with page after page of messages but with a sense of urgency suggesting the recipient has nothing better to do than answer them.

Worse, it is so simple and seemingly cost-free to forward copies of documents that a single carriage return can dispatch fifteen or fifty thousand unwelcome words into your mailbox. That simple act turns e-mail from a personal and conversational medium into wholesale dumping, which is particularly distressing when you are connected over a low-bandwidth channel.

One journalist commissioned to write about the newcomers and their inconsiderate use of the Internet did his research by sending me and others a four-page questionnaire, without first asking and without the slightest warning. His story should have been a self-portrait.

E-mail can be a terrific medium for reporters. E-mail interviews are both less intrusive and allow for more reflection. I am convinced that e-interviews will become a terrific medium and a standard tool for a large amount of journalism around the world—if only reporters can learn some digital decorum.

The best way to be courteous with e-mail on the Internet is to assume the recipient has a mere 1200 bps and only a few moments of attention. An example of the contrary (a habit practiced to my alarm by all too many of the most seasoned users I know) is returning a full copy of my message with their reply. That is perhaps the laziest way to make e-mail meaningful, and it is a killer if the message is long (and the channel thin).

The opposite extreme is even worse, such as the reply "Sure." Sure, what?

The worst of all digital habits, in my opinion, is the gratuitous copy, the "cc" (who will remember it meant carbon copy?). Mountains of them have scared off many senior executives from being on-line. The big problem with electronic ccs is that they can multiply themselves, because replies are all too frequently sent to the entire cc-list. You can never really tell if somebody accidentally replied to "all," or did not want or know how to do otherwise. If a person is organizing an impromptu international meeting and invites me and fifty other people to attend, the last thing I want to see is fifty detailed travel arrangements and the discussions of such.

As the Bard might have said, Brevity is the soul of e-mail.

EVEN ON SUNDAY

E-mail is a life-style that impacts the way we work and think. One very specific result is that the rhythm of work and play changes. Nine-to-five, five days a week, and two weeks off a year starts to evaporate as the dominant beat to business life. Professional and personal messages start to commingle; Sunday is not so different from Monday.

Some, especially in Europe and Japan, will say this is a disaster. They wish to leave their work at the office. I certainly don't begrudge people the right to distance themselves from their work. On the other hand, some of us like to be "wired" all the time. It is a simple trade-off. Personally, I'd rather answer e-mail on Sunday and be in my pajamas longer on Monday.

BEING HOME AND ABROAD
AT THE SAME TIME

There is a very good, now quite famous, cartoon of two dogs using the Internet. One dog types to the other: "On the Internet, nobody knows you're a dog." It should be appended with: "And they don't know where you are."

When I fly from New York to Tokyo, roughly fourteen hours, I will type most of the trip and, among other things, compose forty or fifty e-mail messages. Imagine, if upon arrival at my hotel, I handed these to the concierge as faxes and asked that they be sent. Such act would be viewed as a mass mailing. However, when I send them by e-mail, I do it quickly and easily by dialing one local phone number. I am sending them to people, not places. People are sending messages to me, not Tokyo.

E-mail affords extraordinary mobility without anybody having to know your whereabouts. While this may have more relevance to a traveling salesman, the process of staying connected raises some interesting general questions about the difference between bits and atoms in digital life.

When I travel, I make a point of having at least two local telephone numbers that can connect me to the Internet. Contrary to common lore, these are expensive commercial ports, which link me either to that country's local packet system (something I do in Greece, France, Switzerland, and Japan) or to Sprint's or MCI's global packet service. Sprint, for example, has local telephone numbers in thirty-eight Russian cities. Any one of these can connect me to my one-user time-sharing system (the closet) or, as a fallback, to the Media Lab's main computer. From there, I am on the Internet.

Connecting around the globe is a black art. The problem is not being digital, but being plug-ready. Europe has twenty (count 'em) different power plugs! And, while you may have become used to that small little plastic phone jack, the so-called RJ-11 plug, there are 175 others in the world. I am the proud owner of at least one of each and, on a long and circuitous trip, a full 25 percent of my luggage volume is likely to be a combination of phone jacks and power plugs.

Even properly equipped, one can be stymied by the many hotels and almost all telephone booths that have no means for direct modem connection at all. For these occasions, a small acoustic coupler can be Velcroed to the handset. This is a task whose difficulty is proportional to the degree of overdesign in the telephone handset.

Once connected, the bits find no problem getting back home, even through the most antiquated, rotary, analog telephone switch, though sometimes this requires very low-speed and highly error-corrected transmission.

Europe has started a Europlug program to come up with a single power plug design that has three goals: 1) it does not look like any of the current plugs, 2) it has the safety features of all existing plugs, and 3) it gives no country an economic advantage (the latter being a special trait of European Union thinking). The point is not just plugs. As we evolve our digital living, more and more the roadblocks are likely to be physical, not electronic.

One example of deliberate digital sabotage is when hotels break off the small plastic release clip of an RJ-11 jack, so that you cannot plug your laptop into the wall. This is worse than charging for incoming faxes. Tim and Nina Zagat have promised to include a notation in their future hotel guides that tells this, so that the digerati can boycott these establishments and take their digital business elsewhere.

HARD FUN

16

TEACHING DISABLED

When the Media Lab premiered its LEGO/Logo work in 1989, kids, kindergarten through sixth grade from the Hennigan School, demonstrated their projects before a full force of LEGO executives, academics, and the press. A zealous anchorwoman from one of the national TV networks, camera lights ablazing, cornered one child and asked him if this was not just all fun and games. She pressed this eight-year-old for a typical, "cute," sound-bite reply.

The child was obviously shaken. Finally, after her third repetition of the question and after considerable heat from the lights, this sweaty-faced, exasperated child plaintively looked into the camera and said, "Yes, this is fun, but it's hard fun."

Seymour Papert is an expert on "hard fun." Early on he noted that being "good at" languages is an odd concept when you consider that any run-of-the-mill five-year-old will learn German in Germany, Italian in Italy, Japanese in Japan. As we get older, we seem to lose this ability, but we cannot deny we had it in our youth.

Papert proposed that we think about computers in education, literally and metaphorically, as if creating a country called, say, Mathland, where a child will learn math the same way she learns languages. While Mathland is an odd geopolitical concept, it makes perfect computational sense. In fact, modern computer simulation techniques allow the creation of microworlds in which children can playfully explore very sophisticated principles.

At Hennigan, one six-year-old boy in the so-called LEGO/Logo class built a clump of blocks and placed a motor on top. He connected the two wires of the motor to his computer and wrote a one-line program that turned it on and off. When on, the blocks vibrated. He then attached a propeller to the motor, but for some reason mounted it eccentrically (i.e., not centered, maybe by mistake). Now, when he turned on the motor, the blocks vibrated so much, they not only jumped around the table but almost shook themselves apart (solved by "cheating"—not always bad—with a few rubber bands).

He then noticed that if he turned the motor so that the propeller rotated clockwise, the pile of LEGOs would first jerk to the right and then go into random motion. If he turned it on counterclockwise, the pile would first jerk to the left and then go into random motion. Finally, he decided to put photocells underneath his structure and then set the blocks on top of a black squiggly line he had drawn on a large white sheet of paper.

He wrote a more sophisticated program that first turned on the motor (either way). Then, depending on which photocell saw black, it would stop the motor and start it up clockwise, to jerk right, or counterclockwise, to jerk left, thereby getting back onto the line. The result was a moving pile of blocks that followed the black squiggly line.

The child became a hero. Teachers and students alike asked how his invention worked and looked at his project from many different perspectives, asking different questions. This small moment of glory gave him something very important: the joy of learning.

We may be a society with far fewer learning-disabled children and far more teaching-disabled environments than currently perceived. The computer changes this by making us more able to reach children with different learning and cognitive styles.

DON'T DISSECT A FROG, BUILD ONE

Most American children do not know the difference between the Baltics and the Balkans, or who the Visigoths were, or when Louis XIV lived. So what? Why are those so important? Did you know that Reno is west of Los Angeles?

The heavy price paid in countries like France, South Korea, and Japan for shoving many facts into young minds is often to have students more or less dead on arrival when they enter the university system. Over the next four years they feel like marathon runners being asked to go rock climbing at the finish line.

In the 1960s, most pioneers in computers and education advocated a crummy drill-and-practice approach, using computers on a one-on-one basis, in a self-paced fashion, to teach those same God-awful facts more effectively. Now, with the rage of multimedia, we have closet drill-and-practice believers who think they can colonize the pizzazz of a Sega game to squirt a bit more information into the heads of children, with more so-called productivity.

On April 11, 1970, Papert held a symposium at MIT called "Teaching Children Thinking," in which he proposed using computers as engines that children would teach and thus learn by teaching. This astonishingly simple idea simmered for almost fifteen years before it came to life through personal computers. Today, when more than a third of all American homes contain a personal computer, the idea's time has really come.

While a significant part of learning certainly comes from teaching—but good teaching and by good teachers—a major measure comes from exploration, from reinventing the wheel and finding out for oneself. Until the computer, the technology for teaching was limited to audiovisual devices and distance learning by television, which simply amplified the activity of teachers and the passivity of children.

The computer changed this balance radically. All of a sudden, learning by doing became the rule rather than the exception. Since computer simulation of just about anything is now possible, one need not learn about a frog by dissecting it. Instead, children can be asked to design frogs, to build an animal with frog-like behavior, to modify that behavior, to simulate the muscles, to play with the frog.

By playing with information, especially abstract subjects, the material assumes more meaning. I remember when my son's third-grade teacher reported to me sadly that he could not add or subtract a pair of two- or three-digit numbers. How odd, I thought, as he was always the banker when we played Monopoly, and he seemed to do a dandy job at managing those numbers. So I suggested to the teacher that she try posing the same addition as dollars, not just numbers. And, behold, he was suddenly able to add for her three digits and more in his head. The reason is because they were not abstract and meaningless numbers; they were dollars, which related to buying Boardwalk, building hotels, and passing Go.

The computer-controllable LEGO goes one step further. It allows children to endow their physical constructs with behavior. Current work with LEGOs at the Media Lab includes a computer-in-a-brick prototype, which demonstrates a further degree of flexibility and opportunity for Papert's constructivism, and includes interbrick communications and opportunities to explore parallel processing in new ways.

Kids using LEGO/Logo today will learn physical and logical principles you and I learned in college. Anecdotal evidence and careful testing results reveal that this constructivist approach is an extraordinarily rich means of learning, across a wide range of cognitive and behavioral styles. In fact, many children said to have been learning disabled flourish in the constructionist environment.

STREET SMARTS ON THE SUPERHIGHWAY

During the fall break, when I was in boarding school in Switzerland, a number of children including myself could not go home because home was too far away. But we could participate instead in a *concours,* a truly wild goose chase.

The headmaster of the school was a Swiss general (in the reserves, as are most of the Swiss armed forces) and had both cunning and clout. He arranged a five-day chase around the country, where each team of four kids (twelve to sixteen years old) was equipped with 100 Swiss francs ($23.50 at the time) and a five-day railroad pass.

Each team was given different clues and roamed the country, gaining points for achieving goals along the way. These were no mean feats. At one point we had to show up at a certain latitude and longitude in the middle of the night, whereupon a helicopter dropped the next message in the form of a quarter-inch tangled audiotape in Urdu, telling us to find a live pig and bring it to a location that would be given at a certain phone number (which we had to determine by a complex number puzzle about the dates when seven obscure events took place, whose last seven digits made up the number to call).

This kind of challenge has always had an enormous appeal to me, and, sorry to brag, my team did win—as I was convinced it would. I was so taken by this experience, I did the same for my son's fourteenth birthday. However, without the American army at my beck and call, I made it a one-day experience in Boston for his class, broken up into teams, with a fixed budget and an unlimited subway pass. I spent weeks planting clues with receptionists, under park benches, and at locations to be determined through telephone number puzzles. As you might probably guess, those who excelled in classwork were not necessarily the winners—in fact, usually the opposite. There has always been a real difference between street smarts and smart smarts.

For example, to get one of the clues in my wild goose chase, you had to solve a crossword puzzle. The smart-smart kids zoomed to the library or called their smart friends. The street-smart kids went up and down the subway asking people for help. Not only did they get the answers more quickly, but they did so while moving from A to B and gaining distance and points in the game.

Today kids are getting the opportunity to be street smart on the Internet, where *children are heard and not seen.* Ironically, reading and writing will benefit. Children will read and write on the Internet to communicate, not just to complete some abstract and artificial exercise. What I am advocating should not be construed as anti-intellectual or as a disdain for abstract reasonings—it is quite the opposite. The Internet provides a new medium for reaching out to find knowledge and meaning.

A mild insomniac, I often wake up around 3:00 a.m., log in for an hour, and then go back to sleep. At one of these drowsy sessions I received a piece of e-mail from a certain Michael Schrag, who introduced himself very politely as a high school sophomore. He asked if he might be able to visit the Media Lab when he was visiting MIT later in the week. I suggested that he sit in the back of the room of my Friday "Bits Are Bits" class, and that we match him with a student guide. I also forwarded a copy of his and my e-mail to two other faculty members who agreed to see him (ironically so: they thought he was the famous columnist Michael Schrage, whose name has an *e* at the end).

When I finally met Michael, his dad was with him. He explained to me that Michael was meeting all sorts of people on the Net and really treated it the way I treated my *concours.* What startled Michael's father was that all sorts of people, Nobel Prize winners and senior executives, seemed to have time for

Michael's questions. The reason is that it is so easy to reply, and (at least for the time being) most people are not drowning in gratuitous e-mail.

Over time, there will be more and more people on the Internet with the time and wisdom for it to become a web of human knowledge and assistance. The 30 million members of the American Association of Retired Persons, for example, constitute a collective experience that is currently untapped. Making just that enormous body of knowledge and wisdom accessible to young minds could close the generation gap with a few keystrokes.

PLAYING TO LEARN

In October 1981 Seymour Papert and I attended an OPEC meeting in Vienna. It was the one at which Sheik Yamani delivered his famous speech about giving a poor man a fishing rod, not fish—teach him how to make a living, not take a handout. In a private meeting with Yamani, he asked us if we knew the difference between a primitive and an uneducated person. We were smart enough to hesitate, giving him the occasion to answer his own question, which he did very eloquently.

The answer was simply that primitive people were not uneducated at all, they simply used different means to convey their knowledge from generation to generation, within a supportive and tightly knit social fabric. By contrast, he explained, an uneducated person is the product of a modern society whose fabric has unraveled and whose system is not supportive.

The great sheik's monologue was itself a primitive version of Papert's constructivist ideas. One thing led to another and both of us ended up spending the next year of our lives working on the use of computers in education in developing countries.

The most complete experiment in this period was in Dakar, Senegal, where two dozen Apple computers with the programming language Logo were introduced into an elementary school. The children from this rural, poor, and underdeveloped west African nation dove into these computers with the same ease and abandon as any child from middle-class, suburban America. The Senegalese children showed no difference in adoption and enthusiasm due to the absence of a mechanistic, electronic, gadget-oriented environment in their normal life. Being white or black, rich or poor, did not have any bearing. All that counted, like learning French in France, was being a child.

Within our own society we are finding evidence of the same phenomenon. Whether it is the demographics of the Internet, the use of Nintendo and Sega, or even the penetration of home computers, the dominant forces are not social or racial or economic but generational. The haves and the have-nots are now the young and the old. Many intellectual movements are distinctly driven by national and ethnic forces, but the digital revolution is not. Its ethos and appeal are as universal as rock music.

Most adults fail to see how children learn with electronic games. The common assumption is that these mesmerizing toys turn kids into twitchy addicts and have even fewer redeeming features than the boob tube. But there is no question that many electronic games teach kids strategies and demand planning skills that they will use later in life. When you were a child, how often did you discuss strategy or rush off to learn something faster than anybody else?

Today a game like Tetris is fully understandable too quickly. All that changes is the speed. We are likely to see members of a Tetris generation who are much better at rapidly packing the trunk of a station wagon, but not much more. As games move to more powerful personal computers, we will see an increase in simulation tools (like the very popular SimCity) and more information-rich games.

Hard fun.

DIGITAL FABLES AND FOIBLES

17

THE CALL OF THE MODEM

I f you were to hire household staff to cook, clean, drive, stoke the fire, and answer the door, can you imagine suggesting that they *not* talk to each other, not see what each other is doing, not coordinate their functions?

By contrast, when these functions are embodied in machines, we are perfectly prepared to isolate each function and allow it to stand alone. Right now, a vacuum cleaner, an automobile, a doorbell, a refrigerator, and a heating system are closed, special-purpose systems whose designers made no effort to have them intercommunicate. The closest we get to coordinated behavior in appliances is embedding digital clocks in a large number of them. We try to synchronize some functions with digital time but for the most part end up with a collection of whimpering machines, whose flashing 12:00 is like a small cry to "please make me just a little bit more intelligent."

Machines need to talk easily to one another in order to better serve people.

Being digital changes the character of the standards for machine-to-machine communications. People used to sit around tables in Geneva and other such places to hammer out (a telling metaphor from the industrial age) world standards for everything from spectrum allocation to telecommunications protocols. Sometimes this takes so long, as in the case of the telephone standard ISDN (integrated services digital network), that it is obsolete by the time it is agreed upon.

The operating assumption and mind-set of standards committees has been that electrical signals are like screw threads. For nuts and bolts to work from country to country, we have to agree upon every critical dimension, not just some of them. If you had the right number of threads per inch or centimeter, the match of nut and bolt would still not work if the diameter was wrong. The mechanical world is very demanding that way.

Bits are more forgiving. They lend themselves to higher-order descriptions and protocols (a term previously reserved for polite society). Protocols can be very specific about how two machines handshake. The term *handshaking* is the technical term for how two machines establish communications, deciding upon variables to be used in their conversation(s).

Just listen to your fax or data modem next time you use it. All that staticky-sounding noise and the beeps are literally the handshaking process. These mating calls are negotiations to find the highest terrain from which they can trade bits, with the greatest common denominator of all variables.

At a yet higher level, we can think of protocol as meta-standards, or languages to be used to negotiate more detailed bit-swapping methods. In multilingual Switzerland the equivalent is being single and riding on a T-bar ski lift with a stranger: the first thing you negotiate with your T-bar partner (if you talk at all) is what language to speak. TVs and toasters will ask each other the same kind of question as a precursor to doing business.

BITTY THINGS

Twenty-five years ago, I was on an advisory committee to evaluate the final designs of the universal product code, the UPC, the now ubiquitous computer-readable emblem of little vertical bars that got President Bush into such trouble when he expressed amazement at the automated supermarket checkout register. The UPC is on cans, boxes, books (defacing this jacket), and everything but fresh vegetables.

This UPC committee's role was to give the seal of approval to the final bar-code design. After judging the finalists (the bull's-eye design was runner-up), we reviewed a handful of lunatic but intriguing suggestions, such as making all food slightly radioactive, in proportion to its cost, so that each checkout counter became a Geiger counter where shoppers paid for the number of rads in their carts. (It is estimated that a normal can of spinach exposes you to a dose of one-tenth of a micro-rad per kilogram per hour; this is one-billionth of a joule per hour, by comparison to the 100,000 joules of chemical energy, which is why Popeye does much better for his strength by eating it.)

This crazy idea had a small germ of wisdom: Why not make each UPC able to radiate data? Or, why not let it be activatable, so that like a child in kindergarten it can raise its hand?

The reason is that this takes power, and consequently UPCs and other small "name plates" tend to be passive. There are solutions, like taking power from light or using so little power that a small battery is usable for years. When this happens in a tiny format, all "things" can be digitally active. For example, every teacup, article of clothing, and (yes) book in your house can say where it is. In the future, the concept of being lost will be as unlikely as being "out of print."

Active labels are an important part of the future, because they bring into the digital fold small members of the inanimate world that are not electrical: teddy bears, allen wrenches, and fruit bowls. In the more immediate future, active labels will be (and are being) used as badges worn by people and animals. What better Christmas present than an active dog or cat collar, so the household pet can never again be lost (or, more precisely, it can get lost, but you will know where it is).

People already wear active badges for security purposes. A novel application is being developed by Olivetti in England. Wearing one of their badges allows the building to know where you are. When you have a call, the phone you're nearest rings. In the future, such devices will not be tacked on with a clip or safety pin but securely attached or woven into your clothes.

WEARABLE MEDIA

Computing corduroy, memory muslin, and solar silk might be the literal fabric of tomorrow's digital dress. Instead of carrying your laptop, wear it. While this may sound outrageous, we are already starting to carry more and more computing and communications equipment on our body.

The wristwatch is the most obvious. It is certain to migrate from a mere timepiece today to a mobile command-and-control center tomorrow. It is worn so naturally that many people sleep with it on.

An all-in-one, wrist-mounted TV, computer, and telephone is no longer the exclusive province of Dick Tracy, Batman, or Captain Kirk. Within the next five years, one of the largest areas of growth in consumer products is likely to be such wearable devices. Timex already offers wireless communications between your PC and its wristwatch. The Timex watch is expected to be so popular that its clever (optical) transmission software will be incorporated in various Microsoft systems.

Our ability to miniaturize will quickly surpass our ability to power these small objects. Power is an area of technology that has moved at a turtle's pace. If the progress in battery technology developed at the same pace as integrated circuits, we would be commuting to work in cars powered by flashlight batteries. Instead, I carry more than ten pounds of batteries when traveling in order to feed my laptop on a long flight. Over time, batteries for laptops have gotten heavier, as notebook computers have acquired more functions and brighter displays. (In 1979 Sony's Typecorder, the first laptop, used only four AA batteries.)

We are likely to see some imaginative solutions to power in wearable computers. Abercrombie & Fitch already markets a safari hat with a solar cell powering a small fan that blows on your forehead. An excellent and newer candidate for power storage is your belt. Take it off and look at the enormous amount of area and volume it consumes. Imagine a faux cowhide belt with a buckle design that allows it to be plugged into the wall to recharge your cellular phone.

As for antennas, the human body itself can be part of them. Also, the form factor of most antennas lends itself to being woven into fabric or worn like a tie. With a little digital help, people's ears could work just as well as "rabbit ears."

The important point is to recognize that the future of digital devices can include some very different shapes and sizes from those that might naturally leap to mind from our current frames (sic) of reference. Computer retailing of equipment and supplies may not be limited to Radio Shack and Staples, but include the likes of Saks and stores that sell products from Nike, Levi's, and Banana Republic. In the further future, computer displays may be sold by the gallon and painted on, CD-ROMs may be edible, and parallel processors may be applied like suntan lotion. Alternately, we might be living in our computers.

BITS AND MORTAR

Educated as an architect, I have found that many valuable concepts of architecture feed directly into computer design, but so far very little in the reverse, aside from populating our environment with smarter devices, in or behind the scenes. Thinking of buildings as enormous electromechanical devices has so far yielded few inspired applications. Even the Starship *Enterprise*'s architectural behavior is limited to sliding doors.

Buildings of the future will be like the backplanes of computers: "smart ready" (a term coined by the AMP Corporation for their Smart House program). Smart ready is a combination of prewiring and ubiquitous connectors for (future) signal sharing among appliances. You can later add processing of one kind or another, for example, to make the acoustic ambience of four walls in your living room sound like Carnegie Hall.

Most examples of "intelligent environments" I have seen are missing the ability to sense human presence. It is the problem of personal computers scaled up: the environment cannot see or sense you. Even the thermostat is reporting the temperature of the wall, not whether you feel hot or cold. Future rooms will know that you just sat down to eat, that you have gone to sleep, just stepped into the shower, took the dog for a walk. A phone would never ring. If you are not there, it won't ring because you are not there. If you are there and your digital butler decides to connect you, the nearest doorknob may say, "Excuse me, Madam," and make the connection.

Some people call this ubiquitous computing, which it is, and some of the same people present it as the opposite of using interface agents, which it is not. These two concepts are one and the same.

The ubiquity of each person's computer presence will be driven by the various and disconnected computer processes in their current lives (airline reservation systems, point-of-sales data, on-line service utilization, metering, messaging). These will be increasingly interconnected. If your early-morning flight to Dallas is delayed, your alarm clock can ring a bit later and the car service automatically notified in accordance with traffic predictions.

Currently absent from most renditions of the home of the future are household robots: a curious turn, because twenty years ago almost any image of the future included a robotic theme. C3PO would make an excellent butler; even the accent is appropriate.

Interest in household robots will swing back, and we can anticipate digital domestics with legs to climb stairs, arms to dust, and hands to carry drinks. For security reasons, a household robot must also be able to bark like a ferocious dog. These concepts are not new. The technology is nearly available. There are probably a hundred thousand people worldwide who would be willing to pay $100,000 for such a robot. That $10 billion market will not go overlooked for long.

GOOD MORNING, TOASTER

If your refrigerator notices that you are out of milk, it can "ask" your car to remind you to pick some up on your way home. Appliances today have all too little computing.

A toaster should not be able to burn toast. It should be able to talk to other appliances. It would really be quite simple to brand your toast in the morning with the closing price of your favorite stock. But first, the toaster needs to be connected to the news.

Your home today probably has more than a hundred microprocessors in it. But they are not unified. The most integrated home system is probably the alarm system and, in some cases, the remote control of lights and small appliances. Coffee makers can be programmed to grind and brew fresh coffee before you wake up. But if you reset your alarm to ring forty-five minutes later than usual, you will wake up to terrible coffee.

The lack of electronic communication among appliances results in, among other things, very primitive and peculiar interfaces in each. For example, as speech becomes the dominant mode of interaction between people and machines, small accessories will also need to talk and listen. However, each one of them cannot be expected to have the full means of producing and understanding spoken language. They must communicate and share such resources.

A centralist model for such sharing is tempting, and some people have suggested information "furnaces" in our basements—a central computer in the home that manages all input and output. I suspect it will not go that way, and the function will be much more distributed among a network of appliances, including one that is a champion at speech recognition and production. If both your refrigerator and your cupboard keep track of your food by reading universal product codes, only one of them needs to know how to interpret them.

The terms "white goods" and "brown goods" are used to differentiate between kitchen-top appliances like toasters and blenders and larger, usually built-in, machines like dishwashers and refrigerators. The classic division between white and brown does not include information appliances, which must change, because white goods and brown goods will increasingly be both information consuming and producing.

The future of *any* appliance is likely to be a stripped-down or puffed-up PC. One reason to move in this direction is to make appliances more friendly, usable, and self-explicating. Just think for a moment about how many machines you have (microwave oven, fax machine, cellular telephone) that have a giant vocabulary of functions (some useless) about which you have not bothered to learn, just because it is too hard. Here is where built-in computing can help a great deal, beyond just making sure the microwave oven does not soften the Brie into a puddle. Appliances should be good instructors.

The notion of an instruction manual is obsolete. The fact that computer hardware and software manufacturers ship them with product is nothing short of perverse. The best instructor on how to use a machine is the machine itself. It knows what you are doing, what you have just done, and can even guess at what you are about to do. Folding that awareness into a knowledge of its own operations is a small step for computer science, but a giant step forward and away from a printed manual you can never find and rarely understand.

Add some familiarity with you (you are left-handed, hard of hearing, and have little patience with mechanical things), and that machine can be a far better aide (the *e* in *aide* is purposeful) to its own operations and maintenance than any document. Appliances of tomorrow should come with no printed instructions whatsoever (except This Side Up). The "warranty" should be sent electronically by the appliance itself, once it feels it has been satisfactorily installed.

SMART CARS

The cost of the electronics in a modern car now exceeds the cost of its steel. It already has more than fifty microprocessors in it. That does not mean they have all been used very intelligently. You can be made to feel very foolish when you rent a fancy

European car and realize when you are at the front of a long line for gas that you do not know how to electronically unlock the gas tank.

Cars will have smart radios, energy control, and information displays as the predominant population of digital devices. In addition, automobiles will enjoy another very particular benefit of being digital: they will know where they are.

Recent advances in mapping and tracking make it possible to locate a car's position vis-à-vis a computer model of all roads. The location of every road in the United States can fit on a single CD-ROM. With satellites, loran, or dead reckoning (adding the incremental movements of your car), or with a combination of these tracking techniques, cars can be located to within a few feet. Most people remember James Bond's Aston Martin, in which a computer display in the dashboard situated between him and the passenger seat (hers) showed him a map of where he was and where he was headed. This is now a commercial product, widely accepted and growing in its use. In the United States it was first launched by Oldsmobile in 1994.

There is a small problem, however. Many people, especially older drivers, are very bad at refocusing their eyes rapidly. They find it difficult to go from looking at things at infinity to those at two feet (and back and forth). Worse, some of us need to use reading glasses to see a map, which turns us into Mr. Magoo for driving. A much better way to deliver navigational assistance is by voice.

Since you are not using your ears to drive, they make an ideal channel for telling you when to turn, what to look for, and that if you see such-and-such, you have gone too far. The challenge of exactly how to phrase the directions is difficult (that's why humans do such a lousy job of it). The road is filled with many ambiguities. "Take the next right" is perfectly clear if the turn is several hundred feet or yards away. As you get closer, however, is the "next" right this right or the one after?

Though it is possible to build good, digital, voice-output "back seat drivers," we are not likely to see this concept on the U.S. market too soon. Instead, you will see exactly what James Bond had, right or wrong, safe or not. The reason is ridiculous. If the car talks to you and its mapping data are in error, sending you down a one-way street the wrong way, and you crash, guess who is currently liable? On the other hand, if that happens from reading a map, that is your tough luck. In Europe, where they are more enlightened about liability and litigation, Mercedes-Benz will introduce a talking navigational system this year.

Such navigational systems will not be limited to getting you from A to B. There will be new niche markets for acousta-guides of cities you visit ("on your right is the birthplace of . . .") and for information about food and lodging availability ("booked you a great hotel near Exit 3"). In fact, when your smart car of the future is stolen, it can call you up and tell you exactly where it is. Perhaps it will even sound frightened.

DIGITAL PERSONA

One of the reasons that talking cars have been unpopular is that they've had less personality than a seahorse.

In general, our opinion of a computer's personality is derived from all the things it does badly. On occasion, the reverse may happen. One time I doubled over laughing when my spell-check program looked at my dyslexic-style typo *aslo* and proudly suggested that *asshole* was the correct spelling.

Little by little, computers are taking on personalities. A small but very old example is Hayes Corporation's communications software package, Smartcom, which displays a little telephone with a face. The two eyes look at a list of each step in the connection process and, as the computer completes one step and moves to the next, the eyes follow down the list. The face smiles at the end if the handshaking has been successful, and frowns if it has not.

This is not as frivolous as it sounds. The persona of a machine makes it fun, relaxing, usable, friendly, and less "mechanical" in spirit. Breaking in a new personal computer will become more like house-training a puppy. You will be able to purchase personality modules that include the behavior and style of living of fictitious characters. You will be able to buy a Larry King personality for your newspaper interface. Kids might wish to surf the Net with Dr. Seuss.

I am not suggesting that you be interrupted with knock-knock jokes in the middle of writing an important message. But I am suggesting that the style of interaction can be much richer than the simple clicking sounds, tin voices, or repetitious flashes of error messages. We will see systems with humor, systems that nudge and prod, even ones that are as stern and disciplinarian as a Bavarian nanny.

THE NEW
E-XPRESSIONISTS

18

THE SUNDAY PAINTER REVISITED

A refrigerator with a child's drawing stuck to it is as American as apple pie. We encourage our children to be expressive and make things. Then, suddenly, when they reach age six or seven, we switch gears, leaving them with the impression that art class is as extracurricular as baseball and not nearly as important as, say, English or math. The three Rs are for young men and women who want to be somebody and do something. For the next twenty years we force-feed their left brain like a Strasbourg goose, letting the right shrivel into a pea.

Seymour Papert tells the story of a mid-nineteenth-century surgeon magically transported through time into a modern operating theater. That doctor would not recognize a thing, would not know what to do or how to help. Modern technology would have totally transformed the practice of surgical medicine beyond his recognition. If a mid-nineteenth-century schoolteacher were carried by the same time machine into a present-day classroom, except for minor subject details, that teacher could pick up where his or her late-twentieth-century peer left off. There is little fundamental difference between the way we teach today and the way we did one hundred and fifty years ago. The use of technology is almost at the same level. In fact, according to a recent survey by the U.S. Department of Education, 84 percent of America's teachers consider only one type of information technology absolutely "essential": a photo copier with an adequate paper supply.

Nonetheless, we are finally moving away from a hard-line mode of teaching, which has catered primarily to compulsive serialist children, toward one that is more porous and draws no clear lines between art and science or right brain and left. When a child uses a computer language like Logo to make a picture on his computer screen, that image is both an artistic and mathematical expression, viewable as either. Even an abstract concept like math can now use concrete components from the visual arts.

Personal computers will make our future adult population simultaneously more mathematically able and more visually literate. Ten years from now, teenagers are likely to enjoy a much richer panorama of options because the pursuit of intellectual achievement will not be tilted so much in favor of the bookworm, but instead cater to a wider range of cognitive styles, learning patterns, and expressive behaviors.

The middle ground between work and play will be enlarged dramatically. The crisp line between love and duty will blur by virtue of a common denominator—being digital. The Sunday painter is a symbol of a new era of opportunity and respect for creative avocations—lifelong making, doing, and expressing. When retired people take up watercolors today, it is like a return to childhood, with very different rewards from those of the intervening years. Tomorrow, people of all ages will find a more harmonious continuum in their lives, because, increasingly, the tools to work with and the toys to play with will be the same. There will be a more common palette for love and duty, for self-expression and group work.

Computer hackers young and old are an excellent example. Their programs are like surrealist paintings, which have both aesthetic qualities and technical excellence. Their work is discussed both in terms of style and content, meaning and performance. The behavior of their computer programs has a new kind of aesthetic. These hackers are the forerunners of the new e-xpressionists.

THE DRAW OF MUSIC

Music has proven to be one of the most important shaping forces in computer science.

Music can be viewed from three very powerful and complementary perspectives. It can be considered from the digital signal-processing point of view—such as the very hard problems of sound separation (like taking the noise of a fallen Coke can out of a music recording). It can be viewed from the perspective of musical cognition—how we interpret the language of music, what constitutes appreciation, and where does emotion come from? Finally, music can be treated as artistic expression and narrative—a story to be told and feelings to be aroused. All three are important in their own right and allow the musical domain to be the perfect intellectual landscape for moving gracefully between technology and expression, science and art, private and public.

If you ask an auditorium filled with computer science students how many of them play a musical instrument, or how many consider themselves to have a serious interest in music, most hands shoot up. The traditional kinship between mathematics and music is manifested strikingly in contemporary computer science and within the hacker community. The Media Lab attracts some of its best computer-science students because of its music.

Childhood avocations like art and music, which are intentionally or unintentionally discouraged by parental and social forces, or else viewed solely as a relief valve to the pressures of scholastic success, could shape the lens through which children see and explore entire bodies of knowledge hitherto presented in one way. I did not like history in school, but I can date almost anything from milestones in art and architecture, versus politics and wars. My son inherited my dyslexia but nevertheless can read wind-surfing and ski magazines avidly, from cover to cover. For some people, music may be the way to study math, learn physics, and understand anthropology.

The flip side is: How do we learn about music? In the nineteenth and early twentieth centuries, playing music in school was common. The technology of recording music curbed that. Only recently have schools started to return to learning music by making it, versus just listening to it. The use of computers to learn music at a very young age is a perfect example of the benefit computers provide by offering a complete range of entry points. The computer does not limit musical access to the gifted child. Musical games, sound data tapes, and the intrinsic manipulability of digital audio are just a few of the many means through which a child can experience music. The visually inclined child may even wish to invent ways to see it.

ART WITH A CAPITAL "E"

Computers and art can bring out the worst in each other when they first meet. One reason is that the signature of the machine can be too strong. It can overpower the intended expression, as occurs so often in holographic art and 3-D movies. Technology can be like a jalapeño pepper in a French sauce. The flavor of the computer can drown the subtler signals of the art.

Not surprisingly, the mutual reinforcement of computers and art has been most effective in music and the performing arts, where the technology of performing, disseminating, and experiencing a work of art most easily commingles. Composers, performers, and audience can all have digital control. If Herbie Hancock released his next piece on the Internet, it would not only be like playing to a theater with 20 million seats in it, but each listener could transform the music depending upon her personal situation. For some this may be as simple as varying the volume. For others it may be turning the music into karaoke. For yet others it may be the modification of the orchestration.

The digital superhighway will turn finished and unalterable art into a thing of the past. The number of mustaches given to Mona Lisa is just child's play. We will see serious digital manipulation performed on said-to-be-complete expressions moving across the Internet, which is not necessarily bad.

We are entering an era when expression can be more participatory and alive. We have the opportunity to distribute and experience rich sensory signals in ways that are different from looking at the page of a book and more accessible than traveling to the Louvre. Artists will come to see the Internet as the world's largest gallery for their expressions and as a means of disseminating them directly to people.

The real opportunity comes from the digital artist providing the hooks for mutation and change. Although this may sound like the total vulgarization of important cultural icons—like turning every Steichen into a postcard or making every Warhol into clip art—the point is, being digital allows the *process,* not just the product, to be conveyed. That process can be the fantasy and ecstasy of one mind, or it can be the collective imagination of many, or it can be the vision of a revolutionary group.

A SALON DES REFUSÉS

The original concept for the Media Lab was to take both human interface and artificial intelligence research in new directions. The new wrinkle was to shape them by the content of information systems, the demands of consumer applications, the nature of artistic thought. The idea was marketed to the broadcasting, publishing, and computer industries as the convergence of the sensory richness of video, the information depth of publishing, and the intrinsic interactivity of computers. Sounds so logical today, but at the time the idea was considered silly. *The New York Times* reported that one *unidentified* senior faculty member thought all the people affiliated with this venture were "charlatans."

The Media Lab is in a building designed by architect I. M. Pei (which Pei designed just after the extension of the National Gallery in Washington, D.C., and just before the pyramid of the Louvre in Paris). It took almost seven years to finance, construct, and assemble the faculty.

As in 1863, when the Paris art establishment declined to let the Impressionists into its official show, the founding faculty members of the Media Lab became a Salon des Refusés and had one of their own, in some cases too radical for their academic department, in some cases too extraneous to their department, and in one case with no department at all. Aside from Jerome Wiesner and myself, the group was composed of a filmmaker, a graphic designer, a composer, a physicist, two mathematicians, and a group of research staff who, among other things, had invented multimedia in the preceding years. We came together in the early 1980s as a counterculture to the establishment of computer science, which at the time was still preoccupied with programming languages, operating systems, network protocols, and system architectures. The common bond was not a discipline, but a belief that computers would dramatically alter and affect the quality of life through their ubiquity, not just in science, but in every aspect of living.

The time was right because personal computers were being born, the user interface was beginning to be seen as central, and the telecommunications industry was being deregulated. Owners and managers of newspapers, magazines, books, movie studios, and television stations were starting to ask themselves what the future might hold. Savvy media moguls, like Steve Ross and Dick Munro of Time Warner, had an intuition about the unfolding digital age. Investing in a lunatic start-up at MIT was an inexpensive hedge. Thus we grew rapidly to three hundred people.

Today, the Media Lab is the establishment. The Internet surfers are the crazy kids on the block. The digerati have moved beyond multimedia into something closer to a real life-style than an intellectual manifesto. Their nuptials are in cyberspace. They call themselves bitniks and cybraians. Their social mobility covers the planet. Today, they are the Salon des Refusés, but their salon is not a café in Paris or an I. M. Pei building in Cambridge. Their salon is somewhere on the Net. It is being digital.

EPILOGUE:
AN AGE OF OPTIMISM

I am optimistic by nature. However, every technology or gift of science has a dark side. Being digital is no exception.

The next decade will see cases of intellectual-property abuse and invasion of our privacy. We will experience digital vandalism, software piracy, and data thievery. Worst of all, we will witness the loss of many jobs to wholly automated systems, which will soon change the white-collar workplace to the same degree that it has already transformed the factory floor. The notion of lifetime employment at one job has already started to disappear.

The radical transformation of the nature of our job markets, as we work less with atoms and more with bits, will happen at just about the same time the 2 billion–strong labor force of India and China starts to come on-line (literally). A self-employed software designer in Peoria will be competing with his or her counterpart in Pohang. A digital typographer in Madrid will do the same with one in Madras. American companies are already outsourcing hardware development and software production to Russia and India, not to find cheap manual labor but to secure a highly skilled intellectual force seemingly prepared to work harder, faster, and in a more disciplined fashion than those in our own country.

As the business world globalizes and the Internet grows, we will start to see a seamless digital workplace. Long before political harmony and long before the GATT talks can reach agreement on the tariff and trade of atoms (the right to sell Evian water in California), bits will be borderless, stored and manipulated with absolutely no respect to geopolitical boundaries. In fact, time zones will probably play a bigger role in our digital future than trade zones. I can imagine some software projects that literally move around the world from east to west on a twenty-four-hour cycle, from person to person or from group to group, one working as the other sleeps. Microsoft will need to add London and Tokyo offices for software development in order to produce on three shifts.

As we move more toward such a digital world, an entire sector of the population will be or feel disenfranchised. When a fifty-year-old steelworker loses his job, unlike his twenty-five-year-old son, he may have no digital resilience at all. When a modern-day secretary loses his job, at least he may be conversant with the digital world and have transferrable skills.

Bits are not edible; in that sense they cannot stop hunger. Computers are not moral; they cannot resolve complex issues like the rights to life and to death. But being digital, nevertheless, does give much cause for optimism. Like a force of nature, the digital age cannot be denied or stopped. It has four very powerful qualities that will result in its ultimate triumph: decentralizing, globalizing, harmonizing, and empowering.

The decentralizing effect of being digital can be felt no more strongly than in commerce and in the computer industry itself. The so-called management information systems (MIS) czar, who used to reign over a glass-enclosed and air-conditioned mausoleum, is an emperor with no clothes, almost extinct. Those who survive are usually doing so because they outrank anybody able to fire them, and the company's board of directors is out of touch or asleep or both.

Thinking Machines Corporation, a great and imaginative supercomputer company started by electrical engineering genius Danny Hillis, disappeared after ten years. In that short space of time it introduced the world to massively parallel computer architectures. Its demise did not occur because of mismanagement or sloppy engineering of their so-called Connection Machine. It vanished because parallelism could be decentralized; the very same kind of massively parallel architectures have suddenly become possible by threading together low-cost, mass-produced personal computers.

While this was not good news for Thinking Machines, it is an important message to all of us, both literally and metaphorically. It means the enterprise of the future can meet its computer needs in a new and scalable way by populating its organization with personal computers that, when needed, can work in unison to crunch on computationally intensive problems. Computers will literally work both for individuals and for groups. I see the same decentralized mind-set growing in our society, driven by young citizenry in the digital world. The traditional centralist view of life will become a thing of the past.

The nation-state itself is subject to tremendous change and globalization. Governments fifty years from now will be both larger and smaller. Europe finds itself dividing itself into smaller ethnic entities while trying to unite economically. The forces of nationalism make it too easy to be cynical and dismiss any broad-stroke attempt at world unification. But in the digital world, previously impossible solutions become viable.

Today, when 20 percent of the world consumes 80 percent of its resources, when a quarter of us have an acceptable standard of living and three-quarters don't, how can this divide possibly come together? While the politicians struggle with the baggage of history, a new generation is emerging from the digital landscape free of many of the old prejudices. These kids are released from the limitation of geographic proximity as the sole basis of friendship, collaboration, play, and neighborhood. Digital technology can be a natural force drawing people into greater world harmony.

The harmonizing effect of being digital is already apparent as previously partitioned disciplines and enterprises find themselves collaborating, not competing. A previously missing common language emerges, allowing people to understand across boundaries. Kids at school today experience the opportunity to look at the same thing from many perspectives. A computer program, for example, can be seen simultaneously as a set of computer instructions or as concrete poetry formed by the indentations in the text of the program. What kids learn very quickly is that to know a program is to know it from many perspectives, not just one.

But more than anything, my optimism comes from the empowering nature of being digital. The access, the mobility, and the ability to effect change are what will make the future so different from the present. The information superhighway may be mostly hype today, but it is an understatement about tomorrow. It will exist beyond people's wildest predictions. As children appropriate a global information resource, and as they discover that only adults need learner's permits, we are bound to find new hope and dignity in places where very little existed before.

My optimism is not fueled by an anticipated invention or discovery. Finding a cure for cancer and AIDS, finding an acceptable way to control population, or inventing a machine that can breathe our air and drink our oceans and excrete unpolluted forms of each are dreams that may or may not come about. Being digital is different. We are not waiting on any invention. It is here. It is now. It is almost genetic in its nature, in that each generation will become more digital than the preceding one.

The control bits of that digital future are more than ever before in the hands of the young. Nothing could make me happier.

ACKNOWLEDGMENTS

In 1976 I wrote a proposal to the National Endowment of the Humanities describing a random access multimedia system that would allow users to hold conversations with famous deceased artists. Dr. Jerome B. Wiesner, then president of MIT, read this cuckoo proposal because the scale of money required his signature. Rather than dismiss it as crazy, he offered to help, realizing that I was wildly outside my depth in, among other things, natural-language processing.

A great friendship began. I started to work with optical videodiscs (very analog at the time). Wiesner pressed for more sophisticated linguistics and deeper commitments to art. By 1979 we talked ourselves, and the MIT Corporation, into building the Media Lab.

The five years which followed found us traveling hundreds of thousands of miles each year, sometimes spending more nights each month with each other than with our families. For me this opportunity to learn from Wiesner and to see the world through his eyes and those of his many brilliant and famous friends was an education. The Media Lab became global because Wiesner was global. The Media Lab valued fine art and rigorous science because Wiesner did.

Wiesner died a month before this book was completed. Until his very last days, he wanted to talk about the issues of "being digital" and his cautious optimism about it. He worried about the misuses to which the Internet would be put as it becomes more widely used, and he worried about unemployment in a digital age that takes jobs and may have fewer to give back. But he always came out on the side of optimism even when optimism concerning his own health and well-being dropped rapidly. His death, on Friday, October 21, 1994, marked the transfer of responsibility to many of us at MIT to do for young people what he had done for us. Jerry, we'll try real hard to follow you.

The Media Lab also started with three other great people to whom I owe special thanks for all they have taught me: Marvin L. Minsky, Seymour A. Papert, and Muriel R. Cooper.

Marvin is the smartest man I know. His humor defies description, and he is arguably the most important computer scientist alive. He is fond of quoting Samuel Goldwyn: "Don't pay any attention to the critics. Don't even ignore them."

Seymour spent his early years with psychologist Jean Piaget in Geneva and shortly afterward became codirector of the AI Lab at MIT with Minsky. He thus brought to the Media Lab a deep understanding of both the human sciences and the sciences of the artificial. As Seymour says, "You cannot think about thinking without thinking about thinking about something."

Muriel Cooper provided the third piece of the puzzle: the arts. She was the primary design force at the Media Lab and took some of the most stable working assumptions of personal computing, like windows, and blew them apart with questions, experiences, and prototype alternatives. Her tragic and unexpected death, on May 26, 1994, ripped a gaping hole in the very being and soul of the Media Lab.

The Media Lab was formed in part out of the earlier Architecture Machine Group (1968–1982), where I did most of my learning with a core of colleagues. I owe enormous thanks to Andy Lippman, who has five patentable ideas per day and from whom many phrases in this book are likely to have come. He knows more about digital television than anyone.

Additional and early insights came from Richard A. Bolt, Walter Bender, and Christopher M. Schmandt, all of whom predate the Media Lab, when we had two small labs, six offices, and a closet. These were the years when we were considered "charlatans" and, as such, they were the golden years. But to be golden, we had to be discovered.

Marvin Denicoff, of the Office of Naval Research, is to computer science what the Medicis were to Renaissance art: he funded people with bold ideas. Because he himself is a playwright, he influenced our research to include interactive cinema years before we would have done so on our own.

When Craig Fields, Denicoff's younger counterpart at ARPA, noticed the extreme lack of American presence in consumer electronics, he took bold measures to advance the idea of a computer TV. Craig's influence was so strong that it cost him his job because at the time his approach flew in the face of the Administration's industry policy (or lack thereof). But during those years he funded most of the research that has led to the field we call multimedia today.

In the early 1980s, we turned to the private sector for support, notably to build what came to be named the Wiesner Building—a $50 million facility. The extraordinary generosity of Armand and Celeste Bartos both started and completed the process of making the Media Lab real. In parallel, we had to make new corporate friends.

The new friends were mostly content providers who had never before interacted with MIT, but who felt (in the early 1980s) their futures were being determined by technology. One exception was Dr. Koji Kobayashi, then chairman and CEO of NEC. Because of his initial support and confidence in the vision of computers and communications, other Japanese companies followed quickly.

In the process of building the seventy-five corporate sponsors we have today, I met many characters—in the best sense of the word. Media Lab students today have an opportunity to schmooze with more CEOs than any other group of students I know. We learn from all such visitors, but three stand out well beyond the others: John Sculley, formerly of Apple Computer; John Evans, CEO of News Electronic Data; and Kazuhiko Nishi, CEO of ASCII Corp.

In addition, I owe special gratitude to Alan Kay of Apple Computer and Robert W. Lucky of Bellcore. Because all three of us are members of the Vanguard Group of CSC, I have developed many of the ideas in this book thanks to their insights. Kay reminds me: "Perspective is worth fifty points of IQ." Lucky was the first to ask, "Is a bit really a bit?"

Labs are not just built on ideas. I owe extreme gratitude to Robert P. Greene, Associate Director for Administration and Finance, who has worked with me for more than twelve years. I can go out on limbs trying new research models and travel incessantly because he is so dedicated and is a man totally trusted by people at the Media Lab and within the MIT administration.

On the teaching front, Stephen A. Benton took an academic organization that was overgrown with weeds and gave it form and character until last July, when he handed that position to his successor, Whitman Richards.

Victoria Vasillopulos runs my office and runs me, both inside and outside MIT, at home and at work. The book suggests that being digital merges home and office, work and play, and it does. Victoria can attest to that. Really intelligent computer agents are far off in time, so having an excellent human one is important (and rare). When I dropped out of sight to finish this book, Victoria's job was to make sure nobody noticed. With the help of her assistants Susan Murphy-Bottari and Felice Napolitano, few did.

The making of the book itself involves a separate story of acknowledgments. Most importantly I want to thank Kathy Robbins, my agent in New York. I met

Kathy more than ten years ago and signed up at the time to be one of her "authors." For the next decade I was so busy building the Media Lab that I could not catch my breath long enough to even think about a book. Kathy had infinite patience and prodded regularly, but ever so gently.

With *Wired* magazine, Louis Rossetto and Jane Metcalfe brilliantly timed their idea of having a life-style magazine for the digital world. My son, Dimitri, was very instrumental in getting me involved, for which I am grateful. I never had written a column before. Some came easy and some were hard. But all were fun and graciously edited by John Battelle. People sent many helpful messages. Raves exceeded rants. All were thoughtful.

When I went to Kathy Robbins with the idea of taking the eighteen *Wired* stories and making them into a book, her reply is best described by the image of a frog's tongue in front of a water bug. Sllllp. Done. Swallowed and signed up in less than twenty-four hours. I was taken to Knopf and introduced to its president, Sonny Mehta, and my editor, Marty Asher. Marty had just discovered America Online (yes, he has two teenagers), and this became our channel for communication. His daughter helped him print from home. Marty became digital very quickly.

Word by word, idea by idea, Marty nursed my dyslexic style into something that is one step away from bullets. For many days, Marty and I were like school kids working all night on a term paper.

Later, Russ Neuman, Gail Banks, Alan Kay, Jerry Rubin, Seymour Papert, Fred Bamber, Michael Schrag, and Mike Hawley all read the manuscript for comment and errors.

Neuman made sure that policy and politics were not off the mark. Banks read the manuscript as a professional reviewer and a professional novice, dog-earing just about every page. Kay found attribution mistakes and balked at places that were too nonsequential, adding the wisdom for which he is so famous. Papert looked at the overall structure and reorganized the beginning. Schrag (sixteen years old) found many errors in the text that the copy editor had missed, notably a typo: 34,800 baud versus 38,400 baud, which nobody would have found! Bamber was the reality check. Rubin was the reader for parliamentary and classical form. Hawley decided to read the book backward, the way he reads music (apparently) to make sure he at least plays the ending of a piece well.

Finally, I must acknowledge my extraordinary parents who gave me infinite amounts of two things beyond love and affection: education and travel. In my day, you had no choice but to move your atoms. By twenty-one, I felt I had seen the world. While this was hardly true, thinking so helped provide me with enough confidence to ignore the critics. For this I am very grateful.

INDEX

bandwidth (*cont'd*)
  economics of, 31–3
  fiber versus copper and, 25–8
  for interactive services, 176
  interlace and, 42
  network configuration and, 33–5
  sensory richness and, 96
banking, electronic, 96
bar codes, 208
battery technology, 210–11
baud, 22
Bell Atlantic, 77, 80
Bellcore, 76
Bender, Walter, 72–3
Benton, Stephen, 123–5
binary representations, 14
bit radiation, 48–50
  licensing, 51–6
bits
  difference between atoms and,
    11–17
  media as, 17–19
  per second (bps), 22, 23, 35
  transporting, 77–9, 177–8
  valuation of, 77
Blockbuster Video, 173–4
books, 69–70
Bosendorfer grand piano, 185–7
both/and interfaces, 97–9
boxels, 123
Brand, Stewart, 169
broadcatching, 169
buildings of future, 211–13

cable television, 34, 45, 79, 80,
    177–8
camcorders, 29, 176
Carnegie Commission, 95
cars, smart, 215–17
cathode ray tubes (CRTs), 44
CBS Records, 81

ccs, electronic, 192
CD-ROM, 6, 64, 67–8, 72
CDs, 185–7
cellular telephones, 52, 94
Clarke, Arthur C., 92–3
CNN, 48, 73, 164
Columbia Pictures, 81
command-and-control systems, 97–8,
    121
commingled bits, 18, *see also* multi-
    media
common sense, 156
CompuServe, 166
computer graphics, 96, 97, 103–15
  consumer, 114–15
  development of, 103–5
  icons in, 108–11
  pixels in, 105–7
  3-D, 120, *see also* virtual reality
  unacceptable jaggies in, 107–8
  windows in, 111–13
constraint resolution, 103
consumer electronics, 80–1, 114–15
copper phone lines, 22, 25–8
copyrights, 58–61
creative expression, 219–26
cross-ownership, 56–8
curtains, 113

DAT (digital audiotape), 58–9
data compression, 15–17, 30–1,
    186
Data General, 46
Dataland, 110
data tablets, 130, 131
decentralization, 157–9, 229
Defense, U.S. Department of (DOD),
    65, 118
depth perception, 117
DirecTV system, 35
Disney, 77, 80

A NOTE ON THE TYPE

This book was produced on a desktop publishing system using a Macintosh computer and QuarkXPress. Final pages were output directly to film.

The text of this book was set in Postscript in ITC Berkeley, originally designed in 1938 by Frederic W. Goudy (1865–1947) for the University of California Press as California Old Style. About 1958, the University agreed to license the sale of the Monotype matrices and the name was changed to Californian. Since the acquisition occurred during the declining days of the Monotype Company in the United States, very few fonts were sold, even though the type was admired as one of Goudy's best book types. In 1983, ITC asked Tony Stan to revise the face for film composition. Leaving the overall appearance unchanged, Stan added three new weights, Medium, Bold, and Black, which permit an extended range of use.

Composed by North Market Street Graphics, Lancaster, Pennsylvania. Printed and bound by Berryville Graphics. Designed by Iris Weinstein.